Grand Solar Minimum Becomes the Ice Age

Science Exploration by Rolf A. F. Witzsche

© Text Copyright Rolf A. F. Witzsche 2018
all rights reserved

This book contains the transcript with images of the exploration video with the above title:
see: http://www.ice-age-ahead-iaa.ca/

Lead in:

The effect of the weakening of the Sun is now so dramatic that one hears a lot of talk about another "Grand Solar Minimum" happening in the near future, such as the Maunder (grand) Minimum that gave us the Little Ice Age.

The concept of a coming "Grand Solar Minimum" is a deception however, because the next Grand Solar Minimum will be the full Ice Age. The underlying support for the Sun no longer exists for a reversal back to 'normal' to be possible from any Grand Minimum. The rapid collapse of the interstellar plasma density for the Solar System, has diminished the background for the powering of the Sun past the point for a possible reversal back to normal. We see the collapse of the solar system by the weakening Sun, evident in the diminishing sunspot numbers, and solar-wind pressure, and in ever-greater volumes of cosmic-ray flux (measured by NASA's spacecraft Ulysses), and larger coronal holes.

The end-result of the weakening Sun promises to be worse than nuclear war, and be more certain if humanity fails to meet the Ice Age challenge by building itself a New World that will enable it to live securely and prosper in an otherwise largely uninhabitable world. If this challenge is taken up, a higher-level paradigm will develop for humanity by which today's terror, conflicts, wars, even nuclear war, and all the grand economic looting, imperial control, and depopulation policies will fall by the wayside as mistakes unworthy of humanity.

The widely expected next Grand Solar Minimum will not end with a recovery, but become the next Ice Age glaciation stage. At this point the Earth becomes a largely uninhabitable 'Ice Planet'. Agriculture ends at this point, and humanity ends with it for the lack of food, unless a new world is created with protected agricultures by technological infrastructures that the Ice Age cannot touch. Here a great crisis is in the making, because the required infrastructures are not being built.

A crisis looms before us, because once the Ice Age glaciation starts, with the collapse of the primer fields for the Sun, it cannot reverse itself without a major build-up in the interstellar plasma stream that is needed for the primer fields to form anew. This build-up will take a long time to develop under the current galactic condition. This means that the recovery of the Sun will take about 90,000 years to happen, long past the anticipated "Grand Solar Minimum" that will become the Ice Age spanning the next 90,000 years.

This book explores numerous historic solar activity measurements. One takes us back in time 150,000 years. This long climate history is preserved in proxy form by Beryllium-10 production ratios that are measurable in the ice of Antarctica. Beryllium-10 is a radio-isotope that is exclusively produced by cosmic-ray flux affecting the Earth atmosphere. It has a half-life of 1.39 million years and is measurable contained in historic ice.

The historic Beryllium-10 measurements drawn from Antarctica reveal some amazing details about the last Ice Age and how it started, and how it may start anew in our time, to which we are already in the boundary zone, where we encounter increasing crop losses, nearly worldwide, that threaten our food supply in the near future.

Table of Contents

- ➤ The Sun emits not only light and heat 10
- Cosmic rays are not rays of energy 11
- Galactic 'cosmic rays' are attenuated 12
- NASA spacecraft Voyager-1 13
- The Sun as the big factor for the large changes 14
- Cosmic-ray particles are 100,000 times smaller 15
- The interaction of cosmic rays 16
- Before all this was recognized 17
- It began with the invention of the telescope 18
- The number and the size of the spots were recorded 19
- Measurable proxy for historic solar activity 20
- The miracle proxy for solar activity 21
- When one of the fast protons manage to collide 22
- The Beryllium-10 atom 23
- The Beryllium values are presented inverted 24
- When the Sun is highly active 25
- By utilizing this principle 26
- Another isotope is Carbon-14 27
- The Carbon-14 isotope measurements 28
- In contrast, Beryllium-10 29
- Beryllium-10 back 140,000 years 30
- Beryllium rates plotted reveals a few surprising details 31
  - ➤ The big cold spell 4700 years ago 32
- Spike in Beryllium 5000 years before the present 33

Greenland ice core at 4770 years before the present 34

The coincidence of the Beryllium spike 35

Colder climate is the result of a weaker Sun 36

Cosmic-ray flux ionizes water vapor 37

> The start of the current interglacial 38

Spikes between 11,000 and 15,000 years be fore the presen 39

The first startup, 15,000 years ago failed 40

The two-step process to get the Sun ramped 41

In the mechanistic universe 42

The two spikes in the Beryllium record 43

> The recovery of the Sun 130,000 years ago 44

The Beryllium spike 130,000 years ago 45

When the solar awakening is complete 46

When the interglacial period ends 47

> cultural effect of cosmic-ray flux 48

High Beryllium level during the glaciation period 49

During interglacial times 50

Beryllium level during the glaciation period 51

The Beryllium spike 4700 years ago, is important 52

The deep-cold period around 4700 years ago 53

A minuscule trail of induced electric currents 54

All the great cultural achievements 55

A similar high-level cosmic-ray background 56

> The Earth's magnetic polarity reversal 57

The famous Beryllium pulse 41,000-years ago 58

The magnetic polarity of the Earth 59

> The Dangaard Oeschger oscillations 60

The researchers Dansgaard and Oeschger 61

Temperature fluctuations in the Greenland ice cores 62

> The solar versus the galactic 63

Not all Beryllium values are solar activity proxies 64

Sun is the major contributor to cosmic-ray flux 65

When the results were plotted 66

When an Earth-oriented coronal hole opens up 67

Sun is by far the major contributor of cosmic-ray flux 68

We have many examples that prove 69

How big a wallop the Sun packs 70

The Sun is evidently the major contributing factor 71

> Determining the start of the next Ice Age 72

The changing Sun is the driving factor for climate changes 73

Professor Dr. Zbigniew Jaworowski, has warned 74

This is what the Beryllium record also indicates 75

The cosmic-ray increase Ulysses has measured 76

Ice Age already beginning, measured in space 77

That the galactic portion is minuscule 78

Living in caves at Pinnacle Point 79

The critical point for us at the present time 80

The beginning level had been 40 times colder 81

It is purely academic at this stage 82

When we get into glacial conditions 83

When the Sun is in its high-powered active mode 84

When the sunspot numbers diminish for a weaker Sun 85

> The Grand Solar Minimum - The Next Ice Age 86

The effect of the weakening of the Sun 87

- The Ice Age will be the next Minimum 88
- During the cold period of the hibernating Sun 89
- During the glacial period, high Beryllium values 90
- We are not at the phase-shift point yet 91
- The Little Ice Age in the 1600s 92
- If the up-lifting of the Sun had not occurred 93
- The warming trend (1715-1998) is reflected 94
- The Sun became up-ramped by a plasma resonance pulse 95
- The warming pulse occurs in long intervals 96
- Also plotted in ratios of Carbon-14 97
- The space race that gave us the Ulysses satellite 98
- Ulysses observed the Sun for 16 years 99
- Suddenly the up-trend was reversed in the 1970s 100
- On-the-ground soil-temperature measurements 101
- Ulysses witnessed the great historic phase shift 102
- The solar wind will cease in the 2030s 103
- same rate of weakening, of the Sun 104
- The end of the solar wind in the 2030s 105
- The recovery of the Sun in the 1700s 106
- The pulses have been getting weaker for 3,000 years 107
- Until a set of the Sun's primer fields collapses 108
- Timing will depend on the resilience of the primer fields 109
- During the startup of the last Ice Age 110
  - ➢ Are we ready? 111
- Are we prepared to live in an uninhabitable world 112
- Society remains choked by its limiting perceptions 113
- Plasma physics, ABOVE mechanistic physics 114

Once the Ice Age Challenge becomes widely recognized 115

   More from the author: 117

14 Libraries of books and video productions... 117

> # The Sun emits not only light and heat

Humanity has been looking at the Sun since ancient times. In some cases the Sun has been worshiped as the giver of life. We can readily feel its warmth. We certainly wouldn't exist without it. But the Sun emits not only light and heat in the form of endless streams of electromagnetic photons. It also emits cosmic rays, which have a large effect on the climate on Earth.

Cosmic rays are not rays of energy

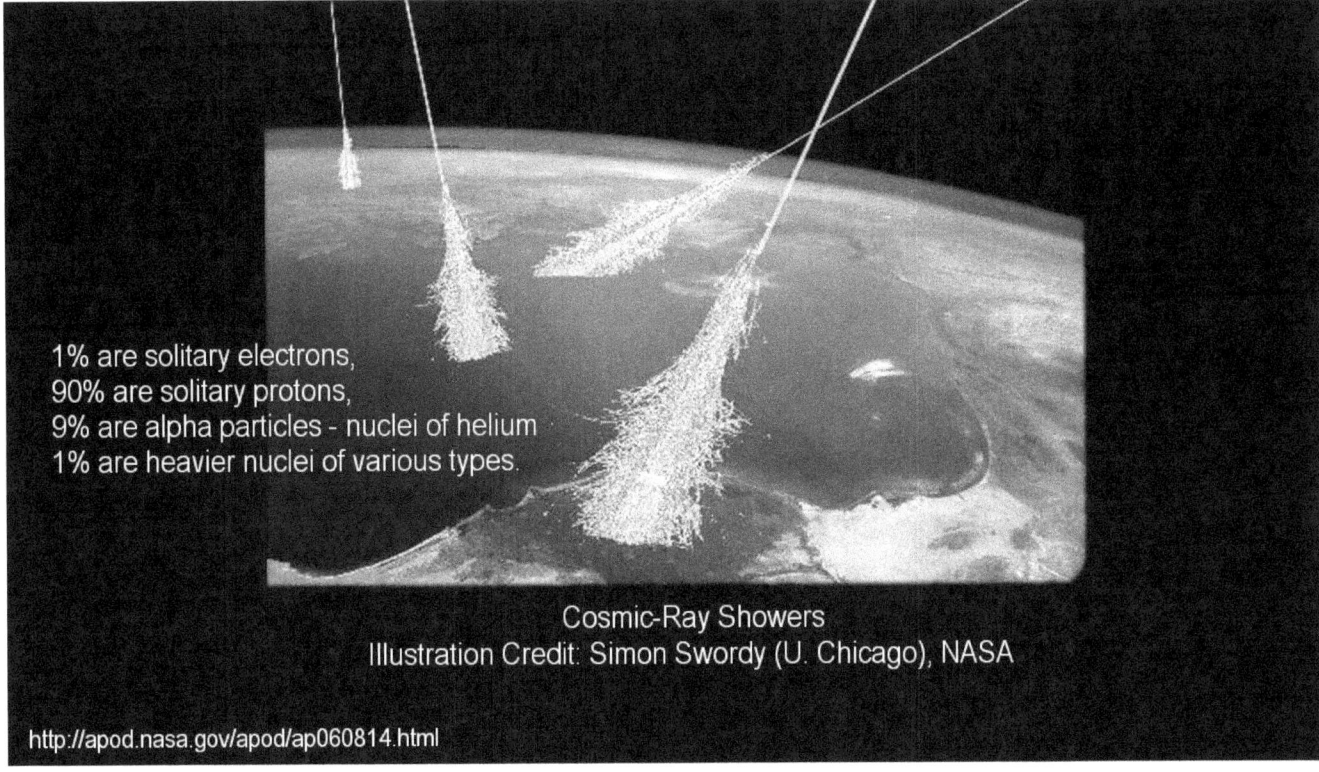

Cosmic rays are not rays of energy in the standard sense. What are termed cosmic rays are single events of fast moving, high-energy plasma particles that carry an electric charge. The majority of the 'cosmic rays' are protons. Most of the 'cosmic ray' events that we encounter on Earth flow from the Sun. A small portion also flow in from galactic space that contains potentially 400 billion suns of a large number of different sizes.

Galactic 'cosmic rays' are attenuated

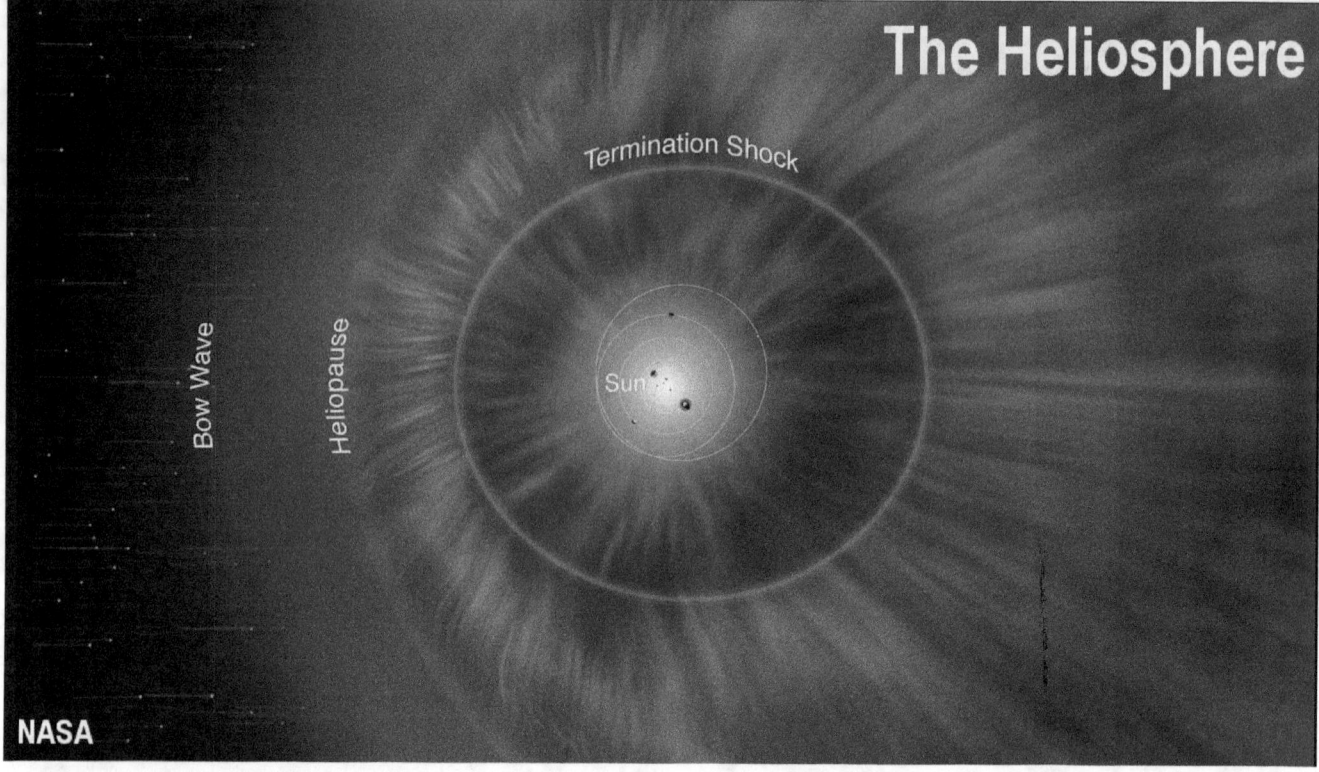

A portion of the galactic 'cosmic rays' are attenuated by a plasma shell that surrounds the solar system, termed the heliosphere, which is formed by the solar wind particles coming to a halt far past the farthest planet. How large a portion of the galactic 'cosmic rays' are blocked by the heliosphere is determined by the strength of it. The attenuation was measured in 2012 as a rather small amount.

NASA spacecraft Voyager-1

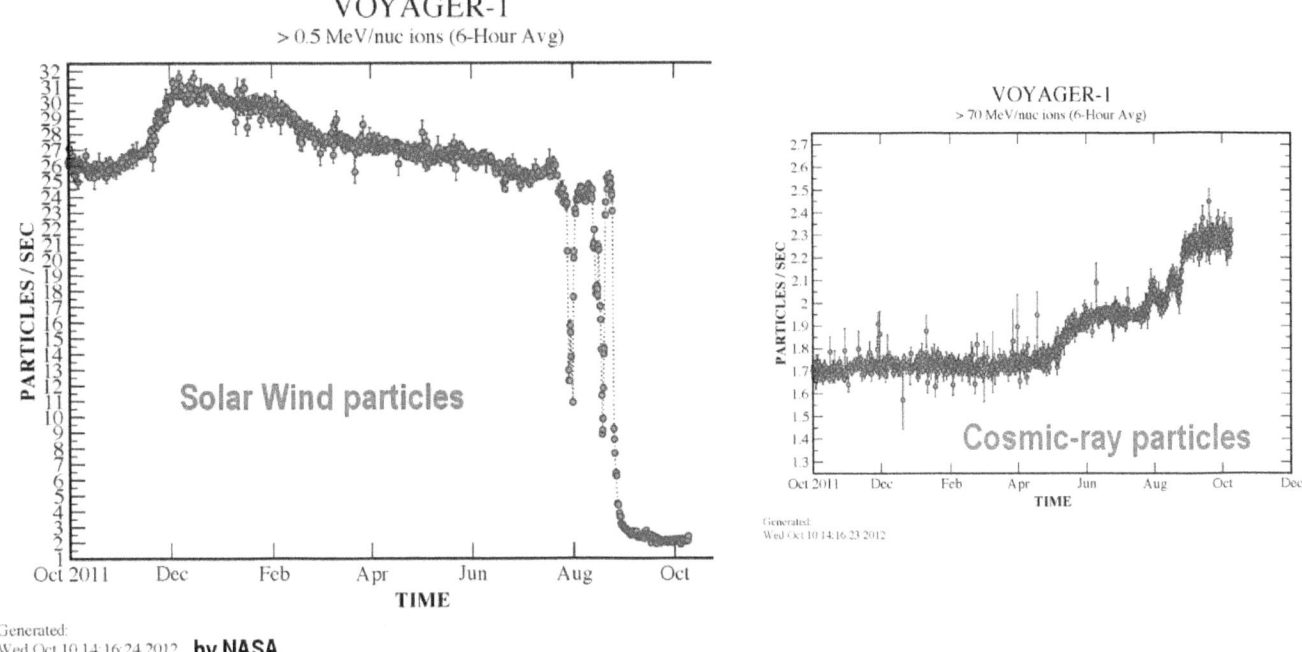

When the NASA spacecraft Voyager-1 crossed into interstellar space in September 2012, it saw the solar wind density diminishing to near zero, as it exited the heliosphere, and it saw the galactic cosmic-ray flux increasing by roughly 35% at this time.

The Sun as the big factor for the large changes

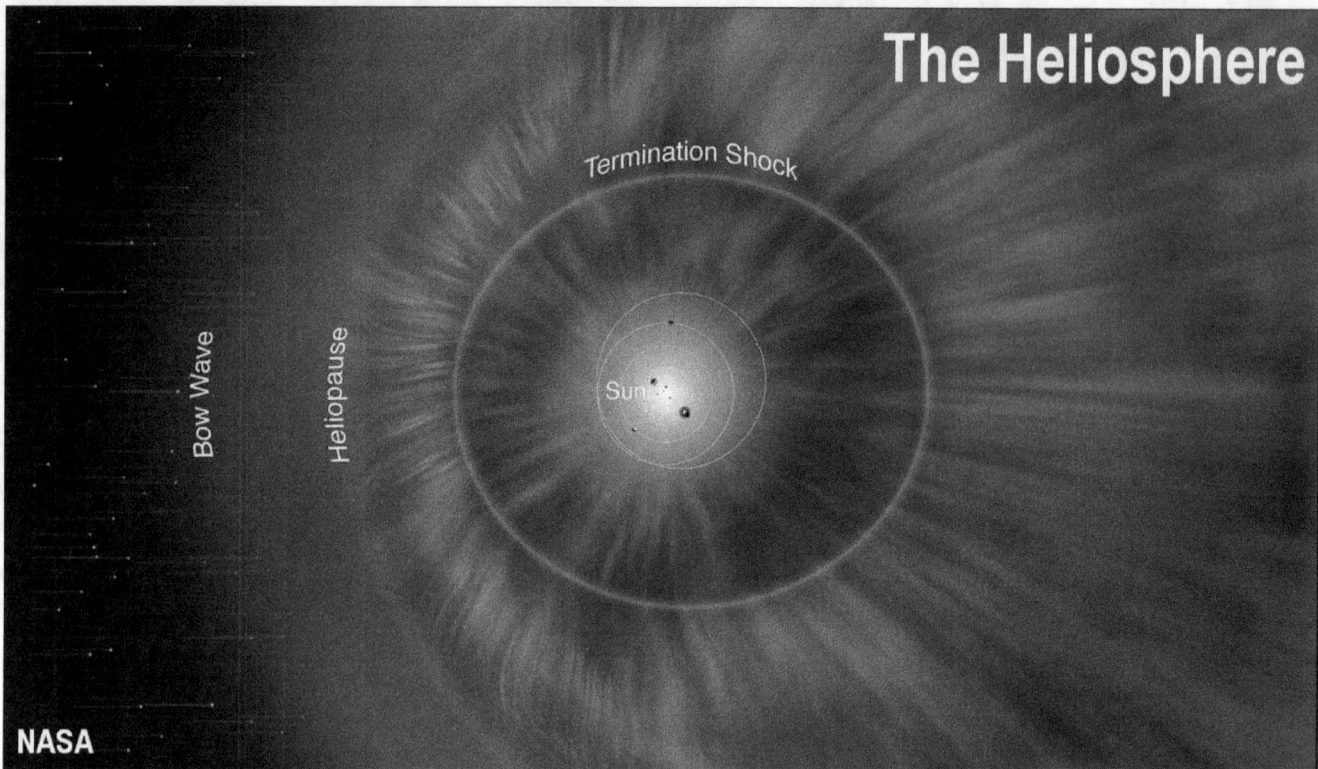

This means that the strength of the heliosphere is not a significant factor for changes in cosmic-ray measurements in the solar system, which leaves us only the Sun as the big factor for the large changes in measured cosmic-ray values that have been measured for historic time on the Earth.

Cosmic-ray particles are 100,000 times smaller

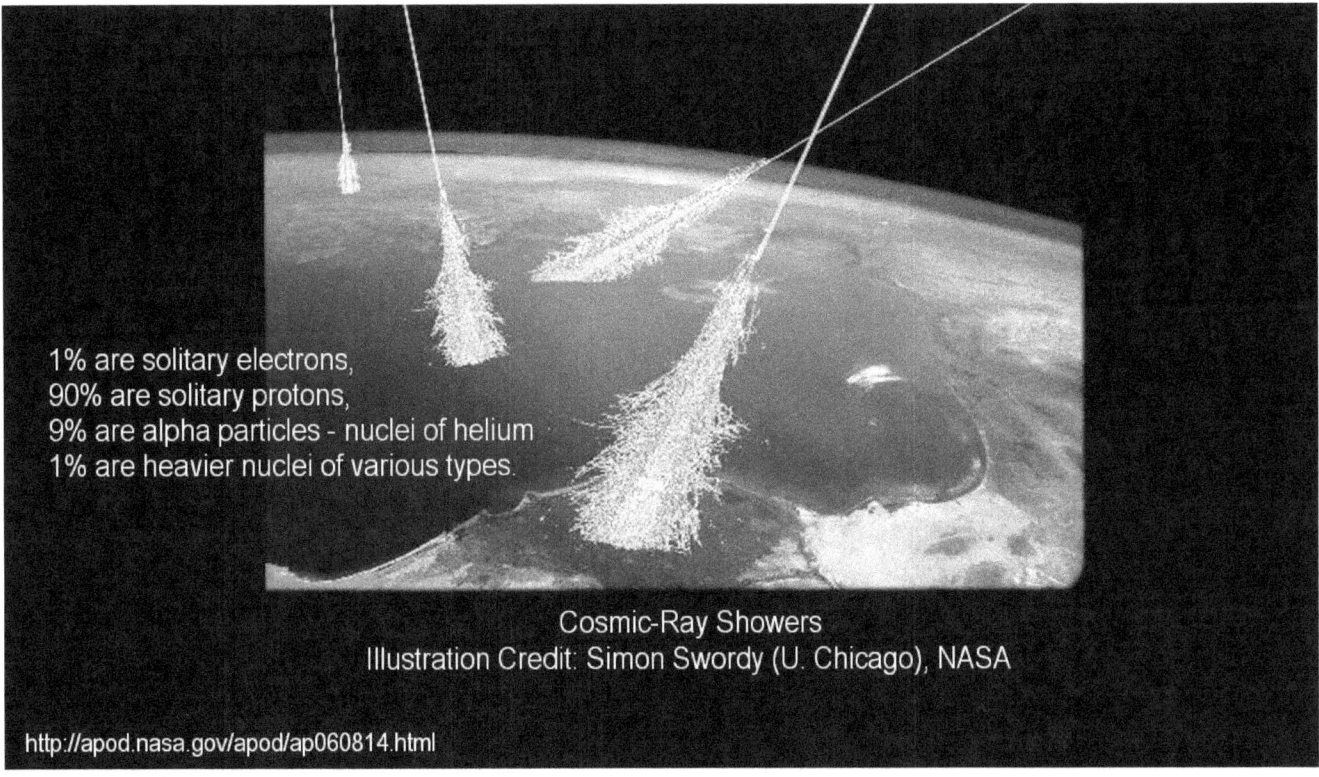

With cosmic-ray particles being essentially plasma particles, they are 100,000 times smaller than the smallest atoms. Cosmic-ray particles are therefore invisible. Nevertheless, their presence can be detected, and their volume be measured by their effect on the Earth's atmosphere.

The interaction of cosmic rays

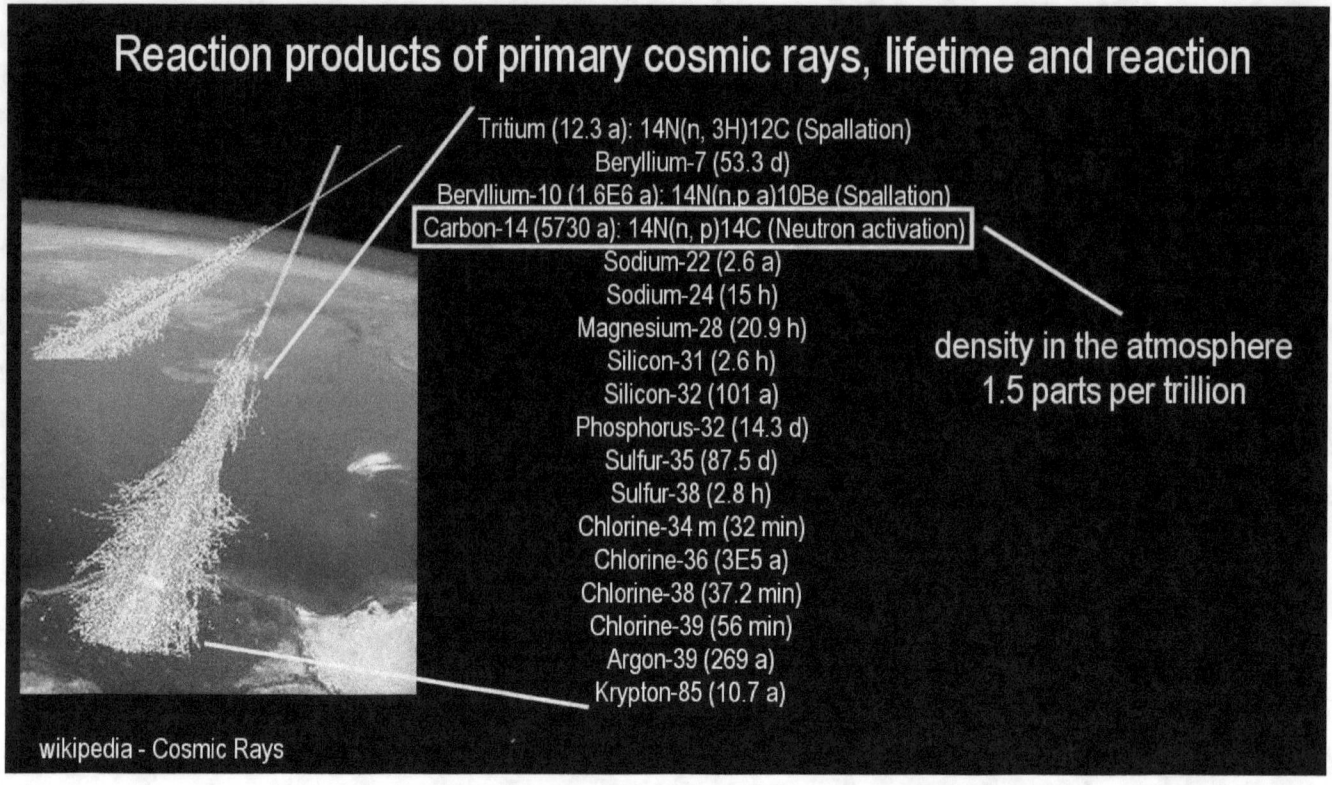

The interaction of cosmic rays with the atmosphere generates radioisotopes that would otherwise not exist. Most decay rapidly, but some persist for thousands of years, even millions of years.

Since most of the cosmic-ray events originate at the Sun, the measurements of the long persistent isotopes gives us the technological means to measure the intensity of the Sun's activity deep into historic times, which is important for climate considerations, especially for Ice Age history considerations.

Before all this was recognized

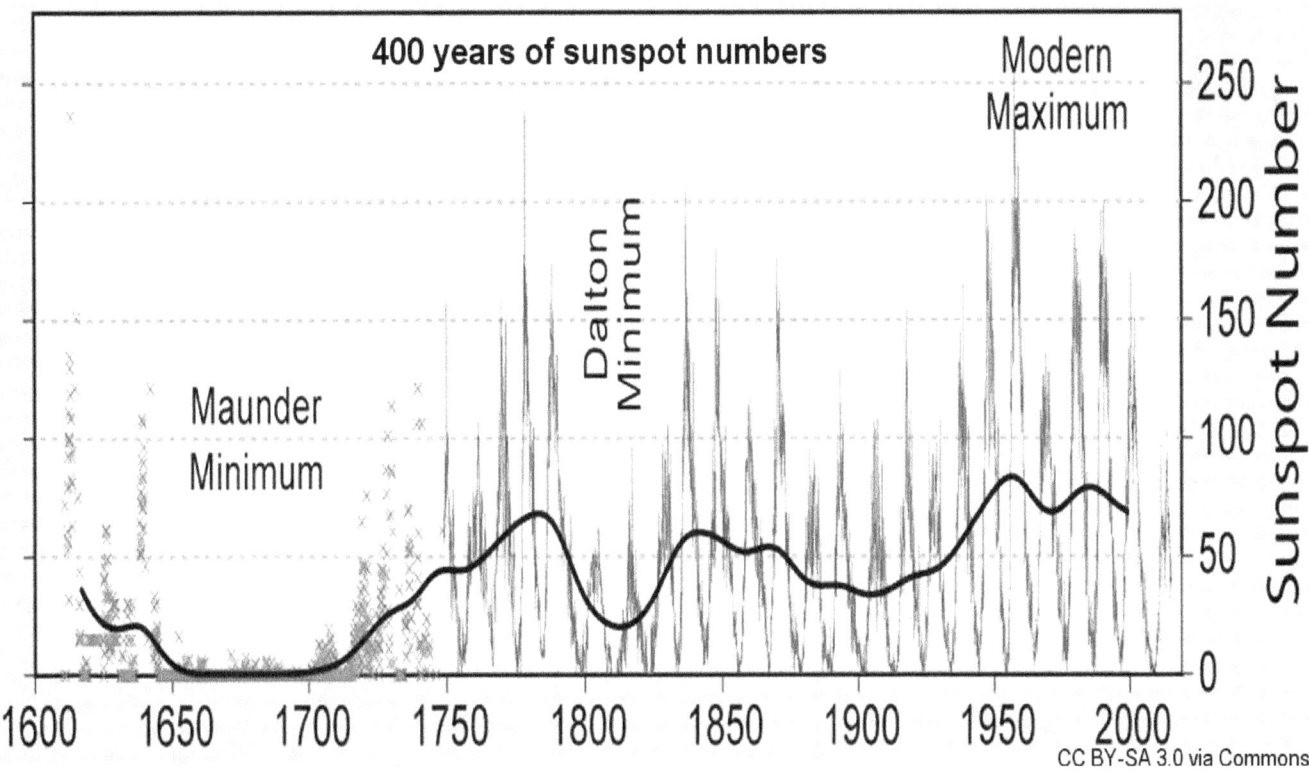

However, before all this was recognized, the changing solar activity over time could only be measured by the changing number of sunspots on the face of the Sun, which reflect changes in solar activity.

It began with the invention of the telescope

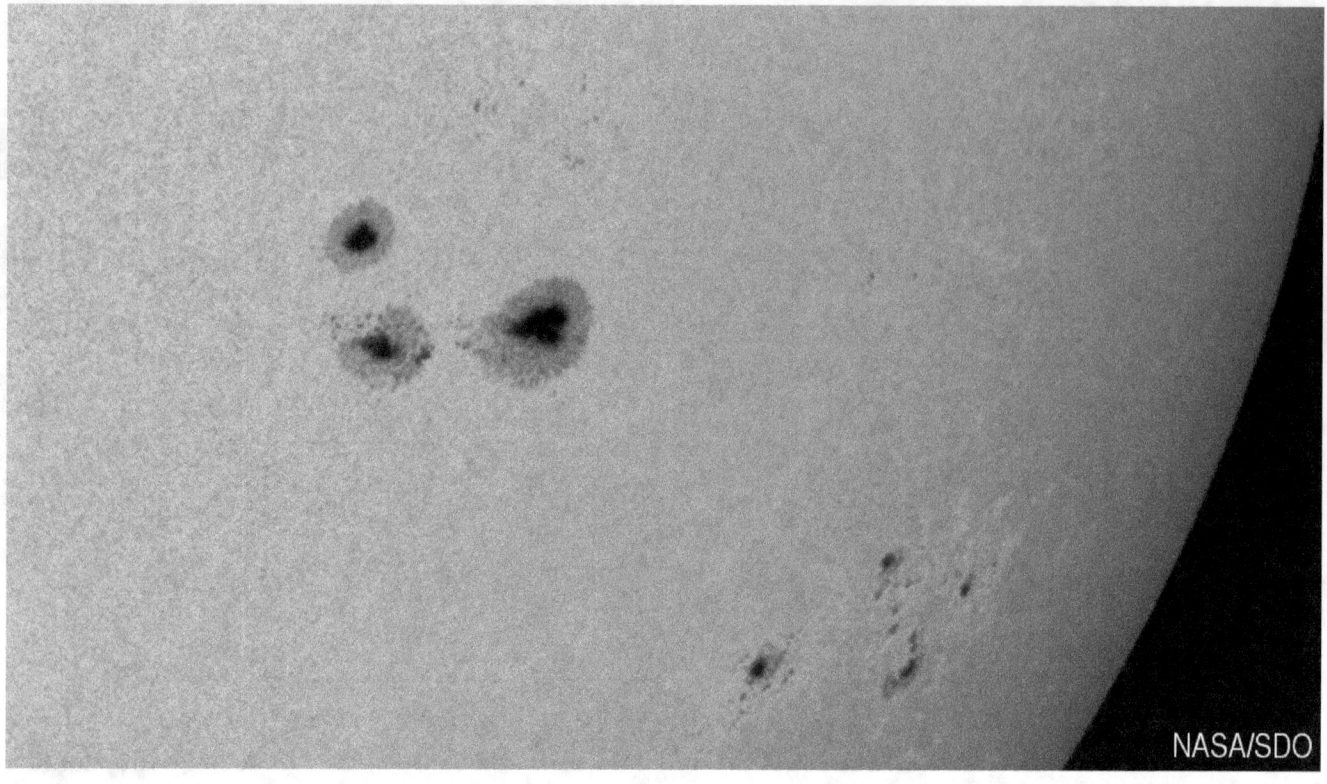

It began with the invention of the telescope that we discovered that the Sun isn't just a scientifically uninteresting ball of fire in the sky. We discovered that things were happening on the face of the Sun. Dark spots were appearing that roused some interest.

The number and the size of the spots were recorded

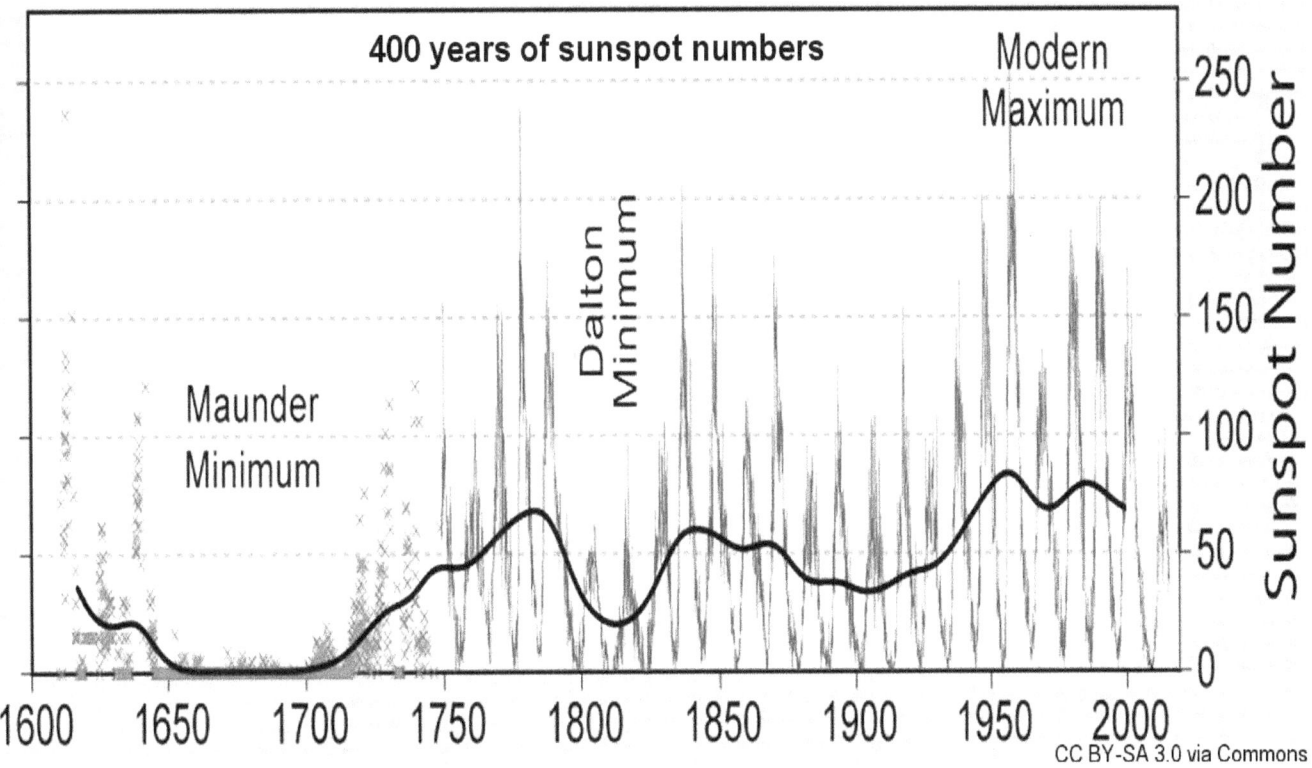

The number and the size of the spots were recorded year after year. Eventually the numbers were plotted, and the discovery was made that during the big minimum of these sunspot numbers, the Maunder Minimum as it was later called, almost no sunspots were recorded for nearly an entire century. Since this was also the time when the Earth had experienced a deep cold spell, the idea emerged that there might be a connection between the century of cold and the century without sunspots. Everyone suffered under the cold. Major rivers froze up. Agriculture suffered severely.

And then in the 1700s the sunspots were back in big numbers, and the numbers were increasing while the Earth was warming up. The sunspots apparently had something to do with the climate. When the numbers were big, the climate was hot. With this the idea dawned that the solar activity that was always changing, had a direct affect on the climate on Earth.

A massive global warming occurred on the Earth in step with the increasing sunspot numbers. The evermore active Sun gave us our wonderfully livable climate back that eventually enabled highly productive agricultures as we have them today.

While the increasing sunspot numbers illustrate plainly that the warming of the Earth from the 1700s to modern time, was caused by the increasing activity of the Sun, we had no way to take this analysis further back into historic time than the time the sunspot numbers had been counted and recorded.

Measurable proxy for historic solar activity

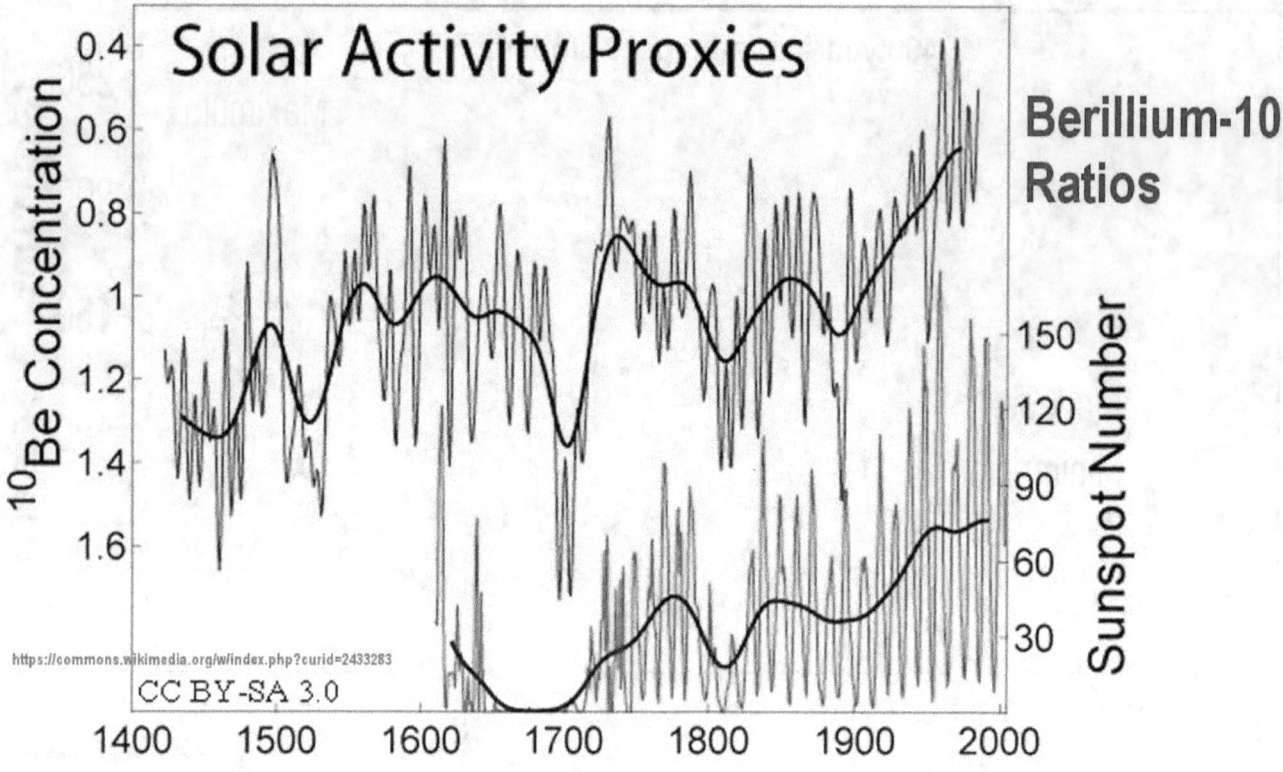

This changed in the late 1970s, when scientists discovered a reliable and measurable proxy for historic solar activity that can take us back in time for millions of years, if need be. And best of all, the proxy agreed with the sunspot numbers for the time that the numbers had been recorded.

The miracle proxy for solar activity

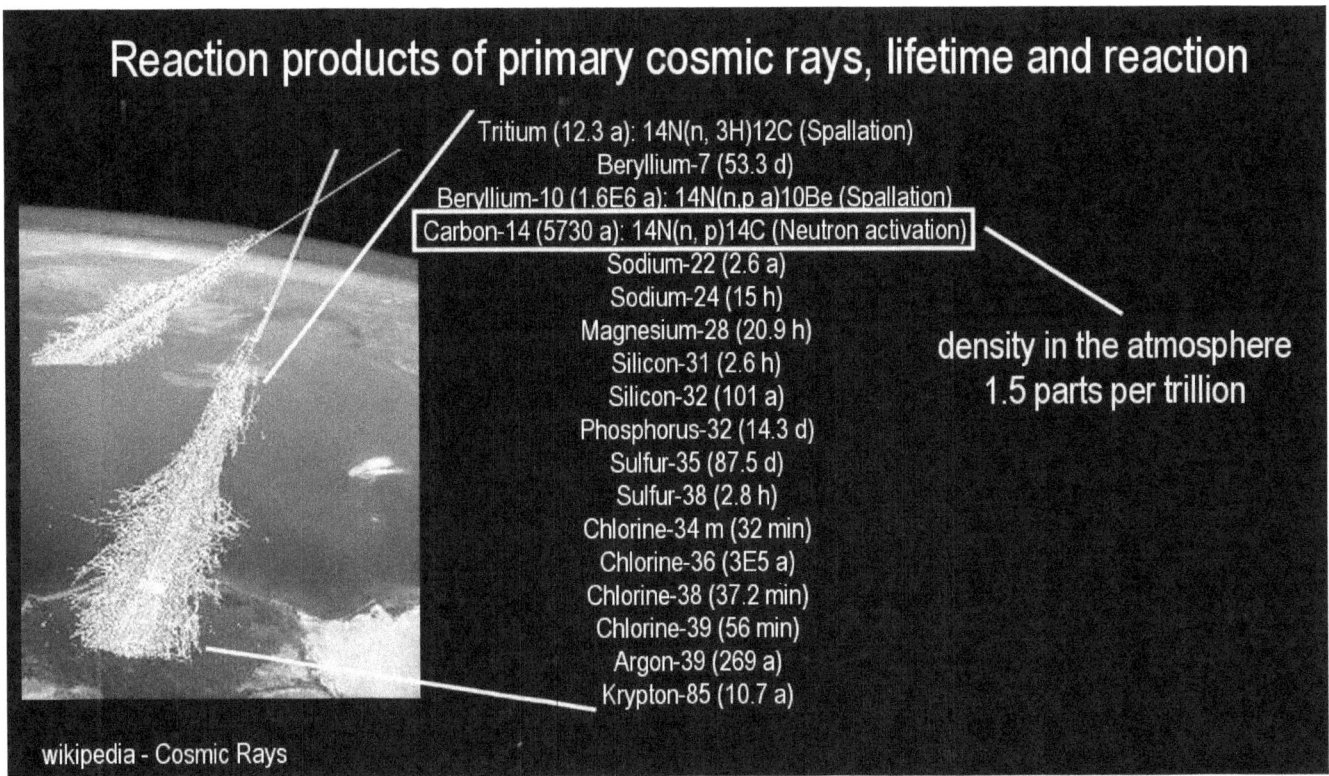

The miracle proxy for solar activity is one of the atomic isotopes that I mentioned earlier, that are created in the upper atmosphere of the Earth by the collision of cosmic-ray protons streaming from the Sun.

When one of the fast protons manage to collide

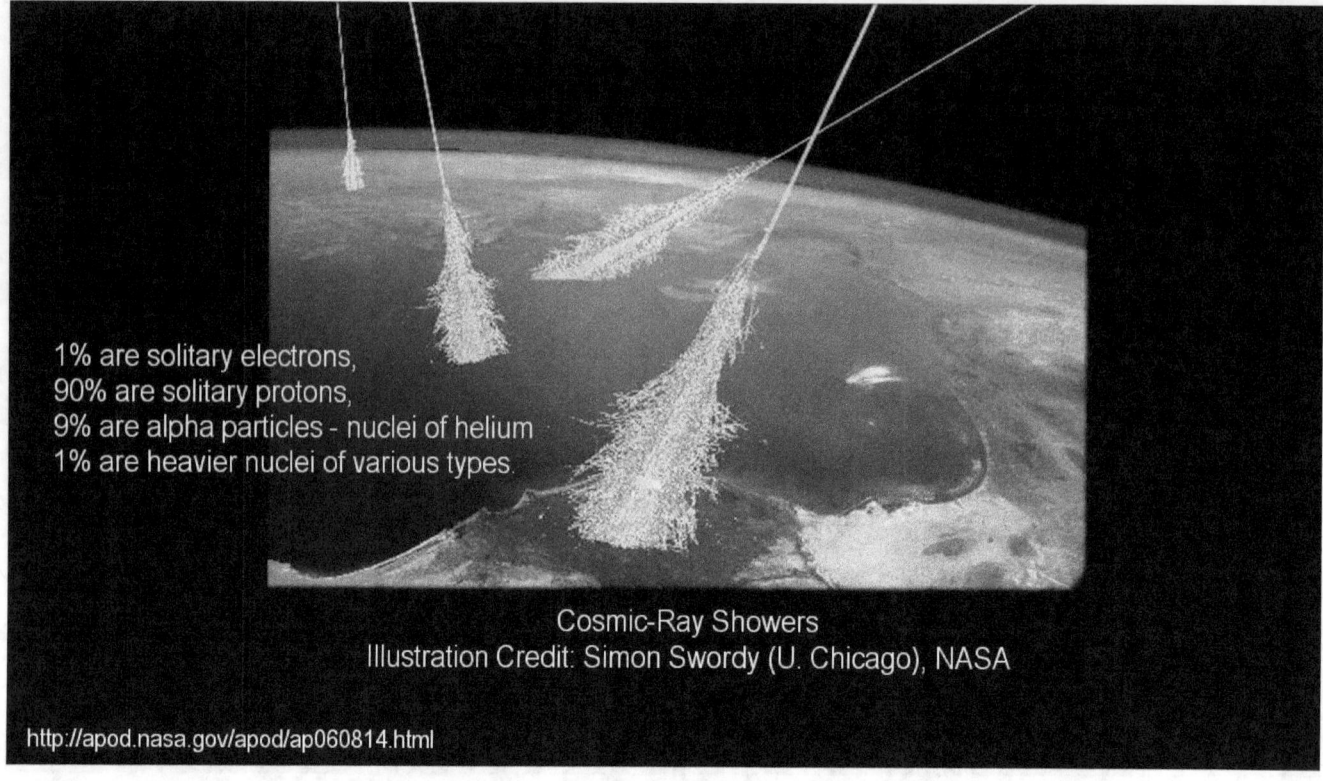

When one of the fast protons manage to collide with either an oxygen atom or a nitrogen atom in the air, the resulting collision breaks the oxygen or nitrogen atom apart into a Berillium-10 atom and some other fragments.

The Beryllium-10 atom

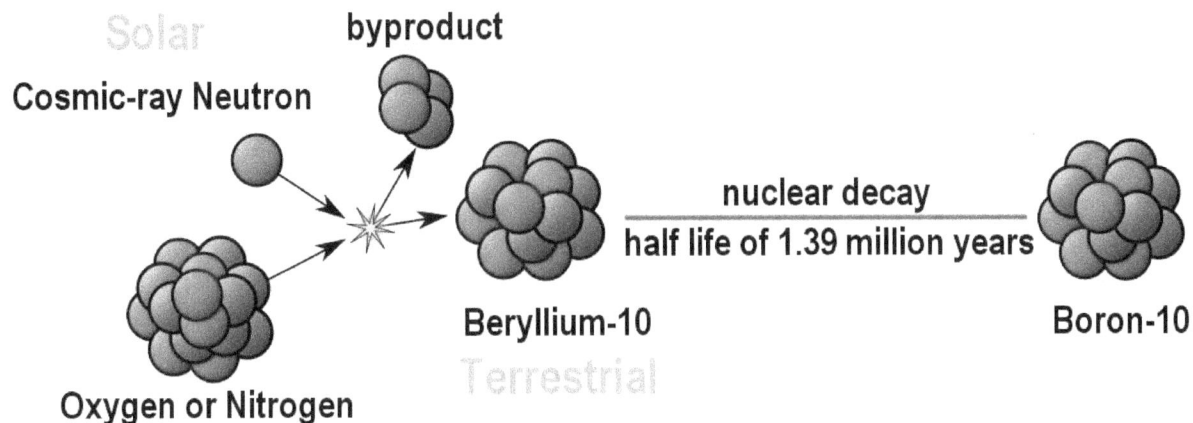

Cosmogenic nuclides production

Oxygen or Nitrogen, Cosmic-ray Neutron, Beryllium-10, Boron-10

The Beryllium-10 atom has a nucleus of 4 protons and 6 neutrons. The imbalance in its atomic nucleus causes the Beryllium-10 atom to decay into Boron-10 that has 5 neutrons and 5 protons in its core. All Boron-10 on the planet has been created by this nuclear decay process. However, this particular radioactive decay takes a long time. Beryllium-10 is the slowest decaying isotope that is created by cosmic-ray collisions in the atmosphere. It has a half life of 1.39 million years.

This means that the volume of Beryllium-10 that we find present in historic samples gives us a measurable proxy for the activity of the Sun at the historic time in which the Beryllium-10 was formed and had settled onto the ground or onto ice sheets, such as in Antarctica. The volume of the Beryllium-10 that has been measured in historic samples provides a measurable value that represents the intensity of the solar activity at that historic time.

The Beryllium values are presented inverted

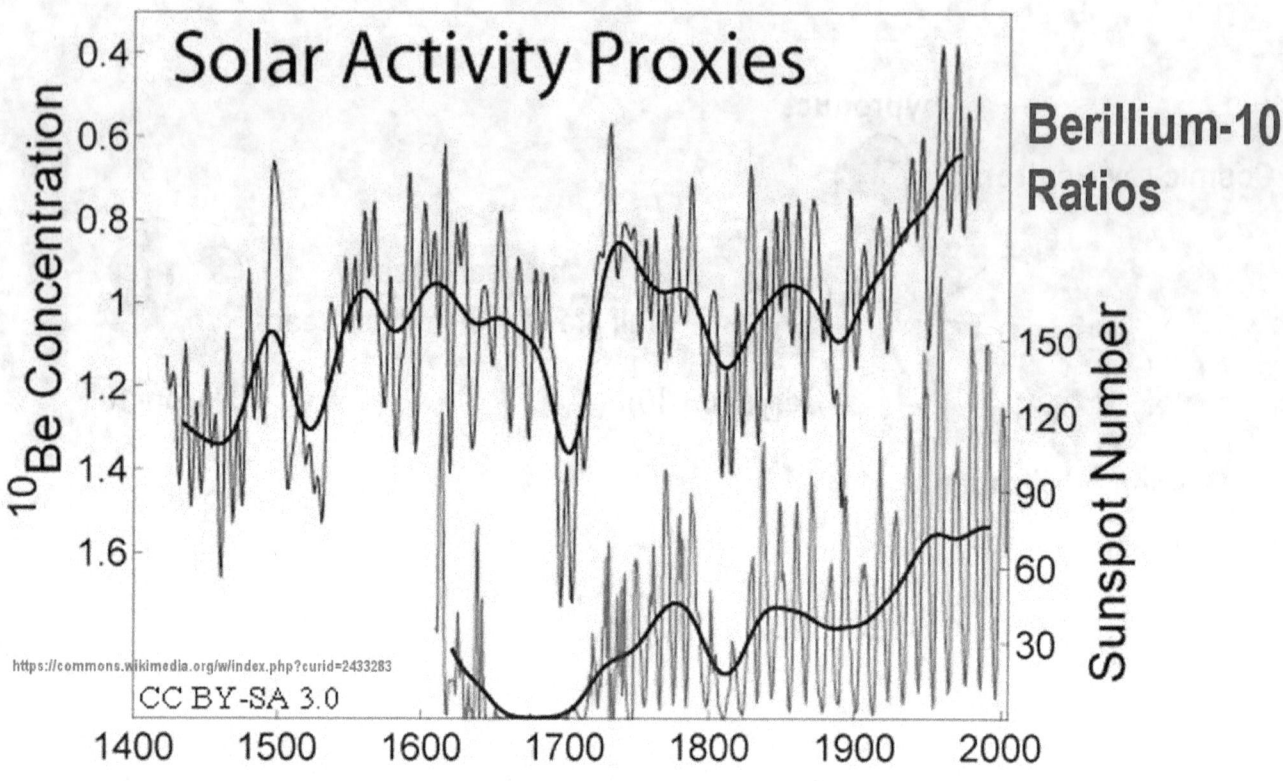

In the case here, the Beryllium values are presented inverted, because when the solar activity is high, less cosmic-ray flux is produced so that the Beryllium measurements are low.

When the Sun is highly active

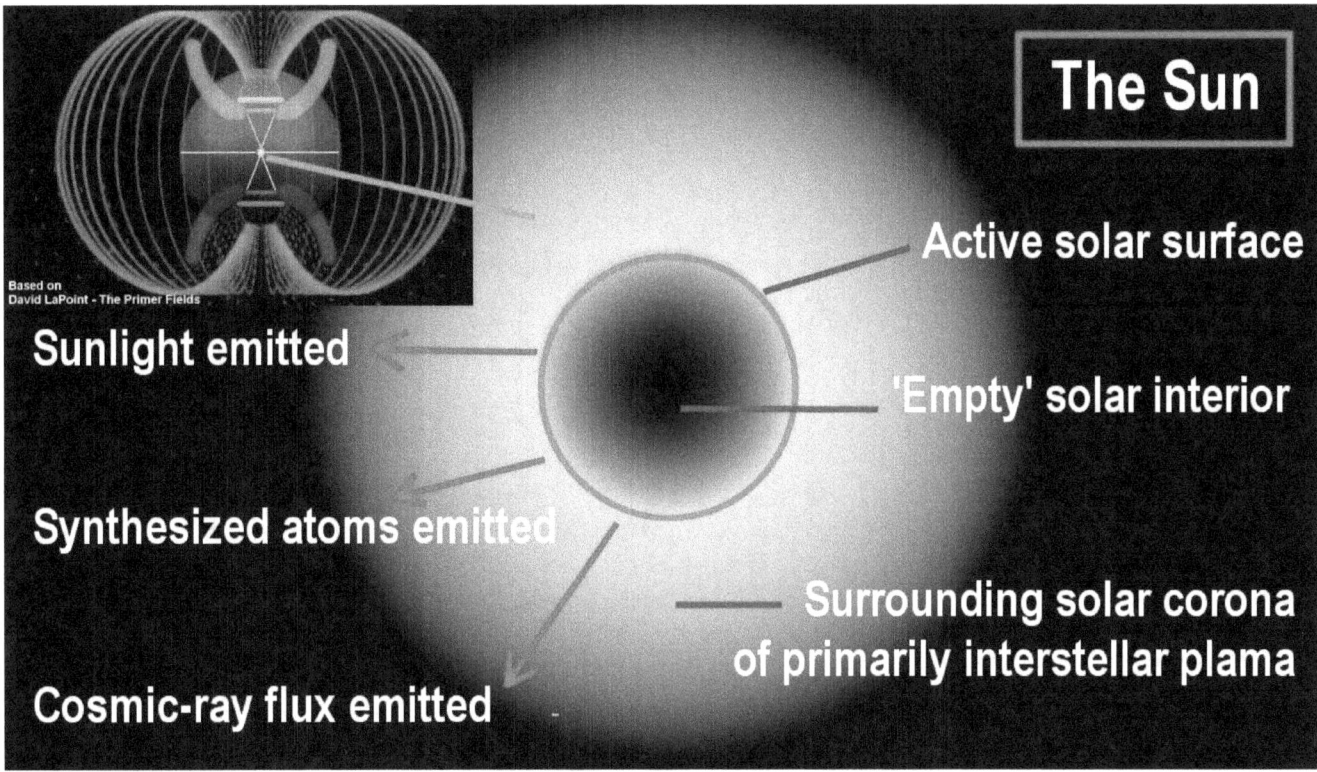

When the Sun is highly active, it is surrounded by a dense sphere of plasma - which is focused onto the Sun from interstellar sources, by which the Sun is also powered. This means that the amount of the Sun's cosmic-ray flux that escapes through the dense plasma, is low, because most of the proton radiation is absorbed in the dense plasma sphere. The result is that fewer cosmic-rays reach the Earth, which results in low Beryllium numbers.

Inversely, when the plasma sphere surrounding the Sun is less dense, which causes the Sun to be less-intensively powered and to be less active, the escaping cosmic ray particles are more abundant. This means that a greater volume of Beryllium-10 is being generated by a weaker Sun.

By utilizing this principle

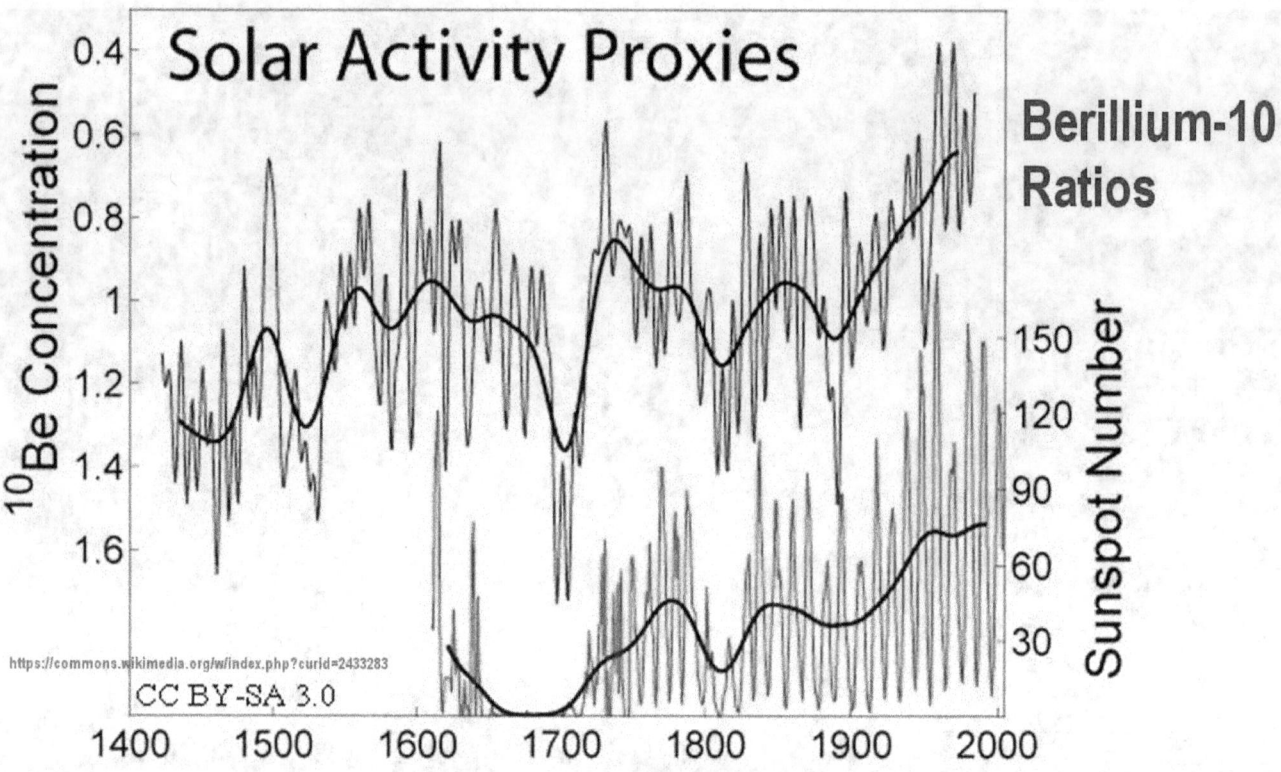

By utilizing this principle, we can determine the strength of the Sun and its solar activity at any point in historic time with an amazingly high resolution. Beryllium takes only 2 years to settle onto the ground. It's resolution is so great that the measured numbers even fluctuate with the solar cycles and accurately match their trend.

Another isotope is Carbon-14

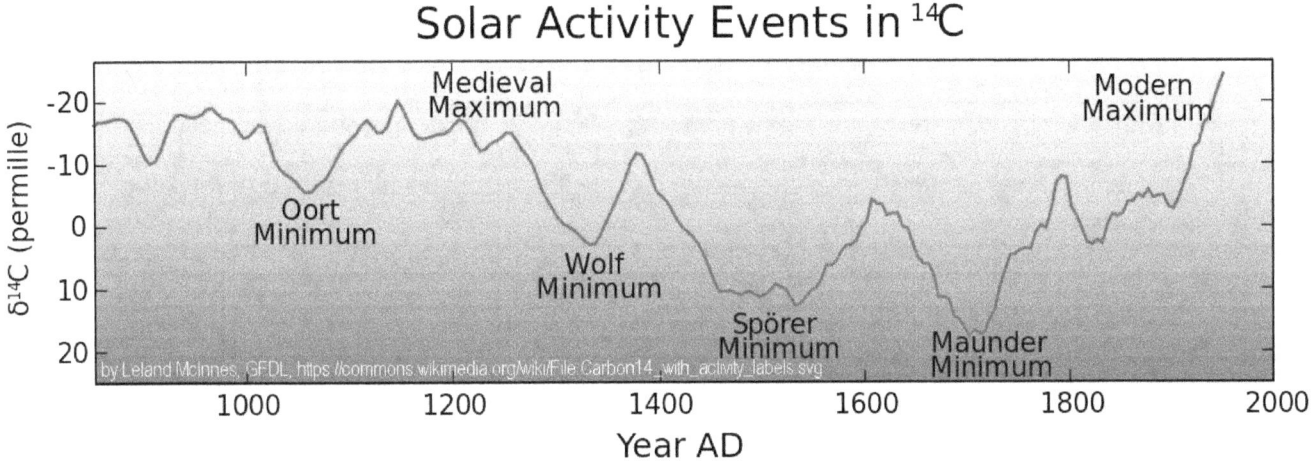

Beryllium-10 is also not the only radio-isotope produced by the Sun. Another isotope is Carbon-14, which likewise reflects the solar activity trend amazingly well though with less resolution in details.

The Carbon-14 isotope measurements

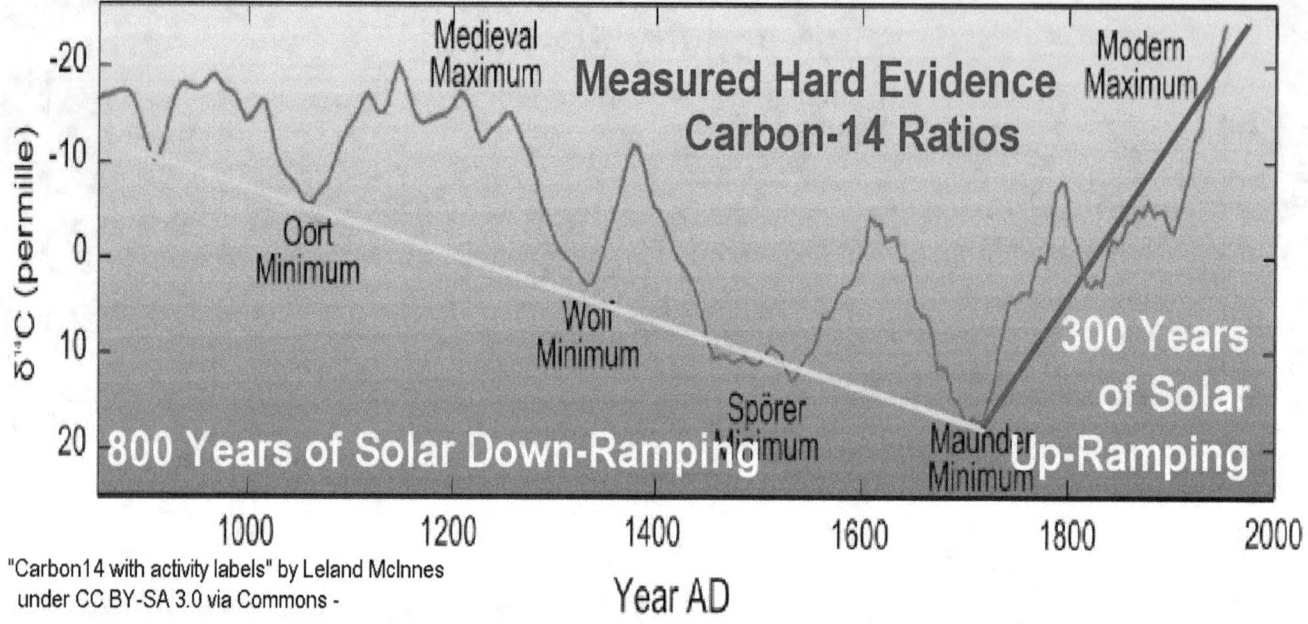

The Carbon-14 isotope measurements show extremely well how the fluctuating solar activity diminished towards the Little Ice Age, and then was ramped up again. The ramping up of the Sun gave us almost 300 years of increasing solar global warming.

Unfortunately, the Carbon-14 isotope is a fast-decaying isotope that has a half-life of only 5,400 years. Its fast decay time makes it unsuitable for looking back in time deep into history.

In contrast, Beryllium-10

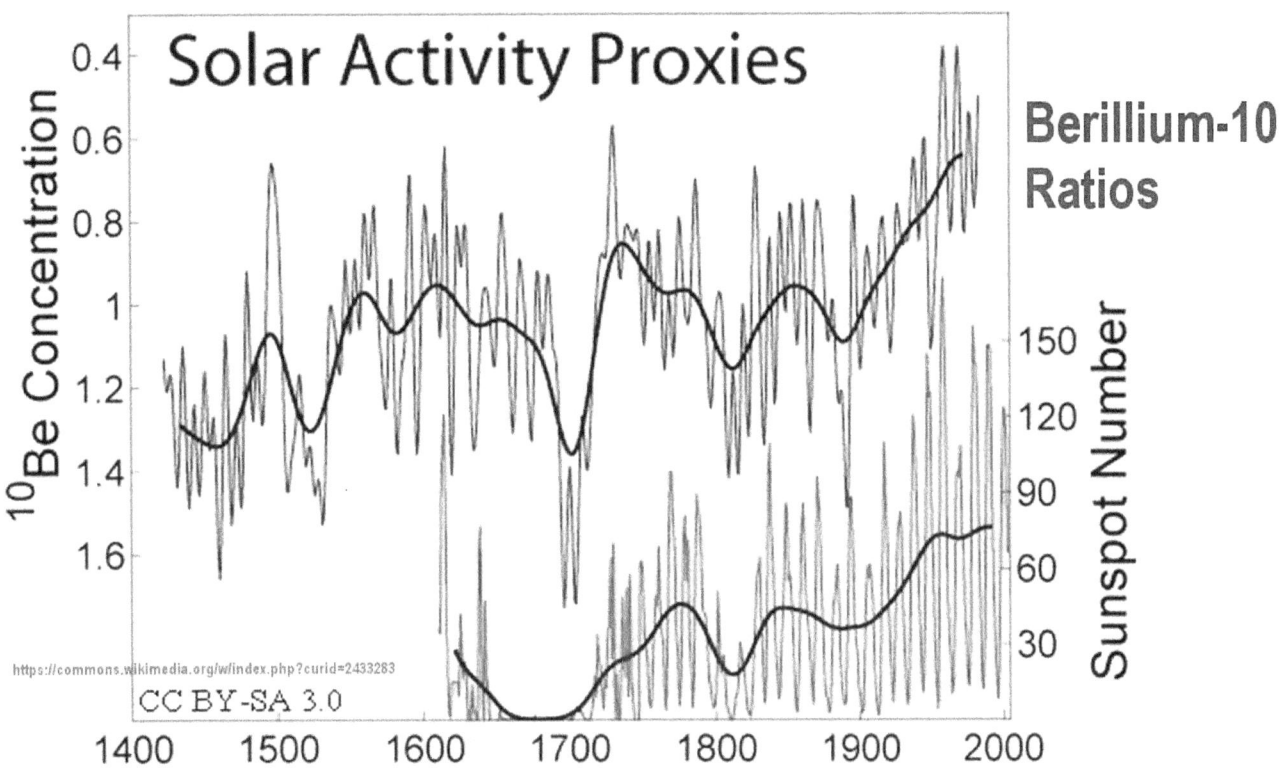

In contrast, Beryllium-10, with its half-life of 1.39 million years enables us to extend our solar exploration deep into historic time, way past the timeframe shown here. With it we can explore the solar activity trends deep into the Ice Ages.

Beryllium-10 back 140,000 years

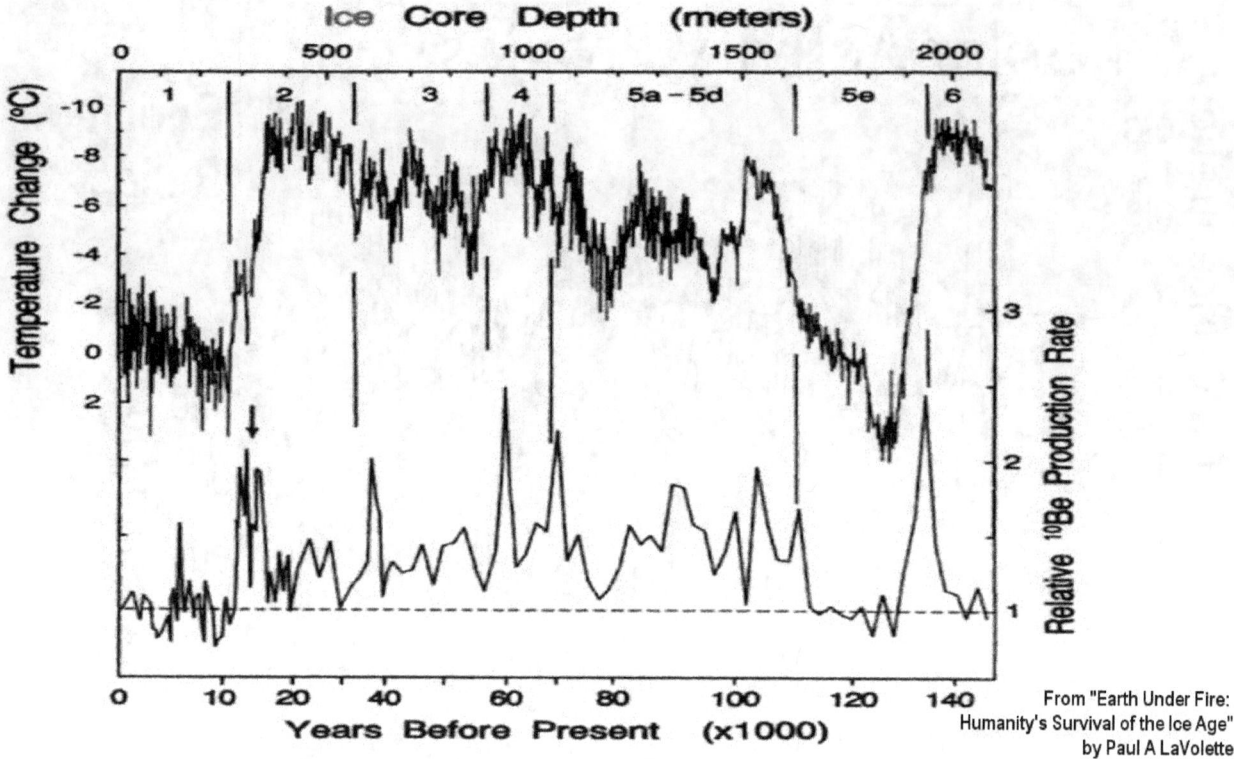

I recently came across a graphic that plots the relative Beryllium-10 production rate all the way back 140,000 years, as it was preserved in the Ice of Antarctica. I found the graphic presented here in the book, 'Earth Under Fire: Humanity's Survival of the Ice Age' by Paul A LaVolette.

Beryllium rates plotted reveals a few surprising details

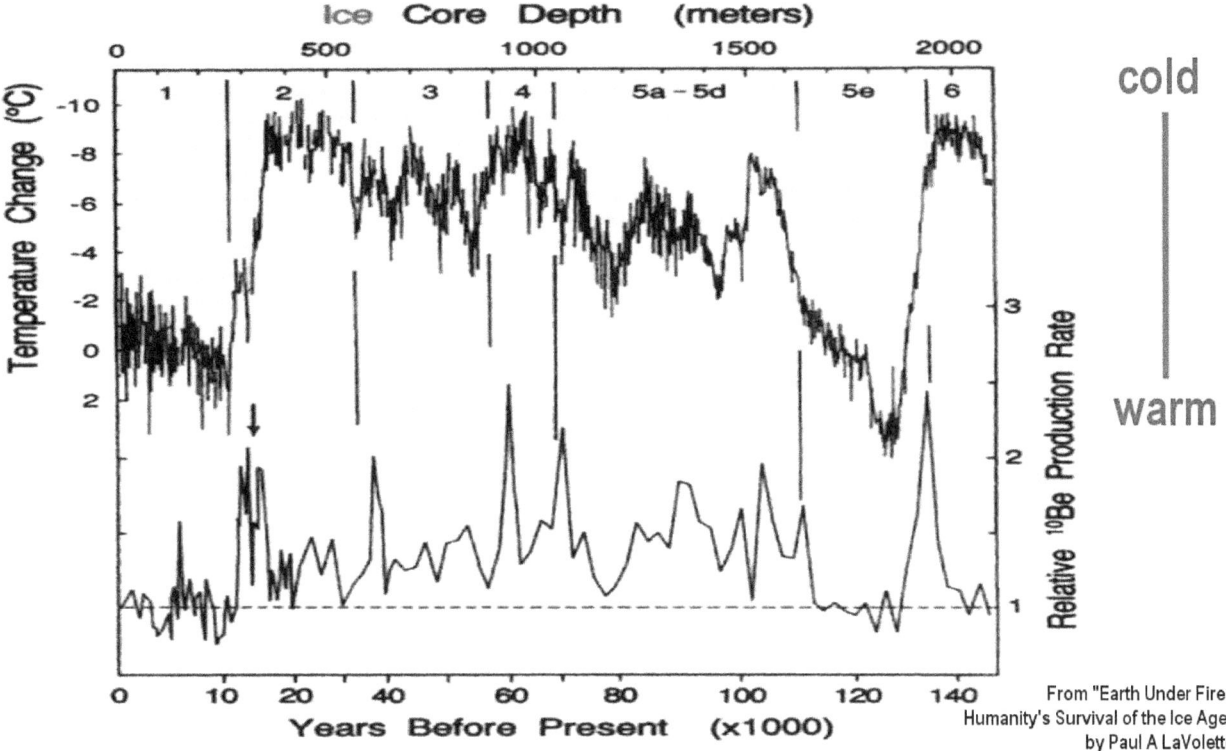

From "Earth Under Fire: Humanity's Survival of the Ice Age" by Paul A LaVolette

In the book, it is the climate record that is plotted inversely rather than the Beryllium record, in order to reflect the inherent inverse relationship of solar cosmic-ray flux and its effect on the climate.

It is rare to see the Beryllium rates plotted over this long timeframe, as presented in Paul LaVolettes book, 'Earth Under Fire.'

The result is interesting, because it reveals a few surprising details.

› The big cold spell 4700 years ago

The big cold spell 4700 years ago

The big cold spell 4700 years ago

Spike in Beryllium 5000 years before the present

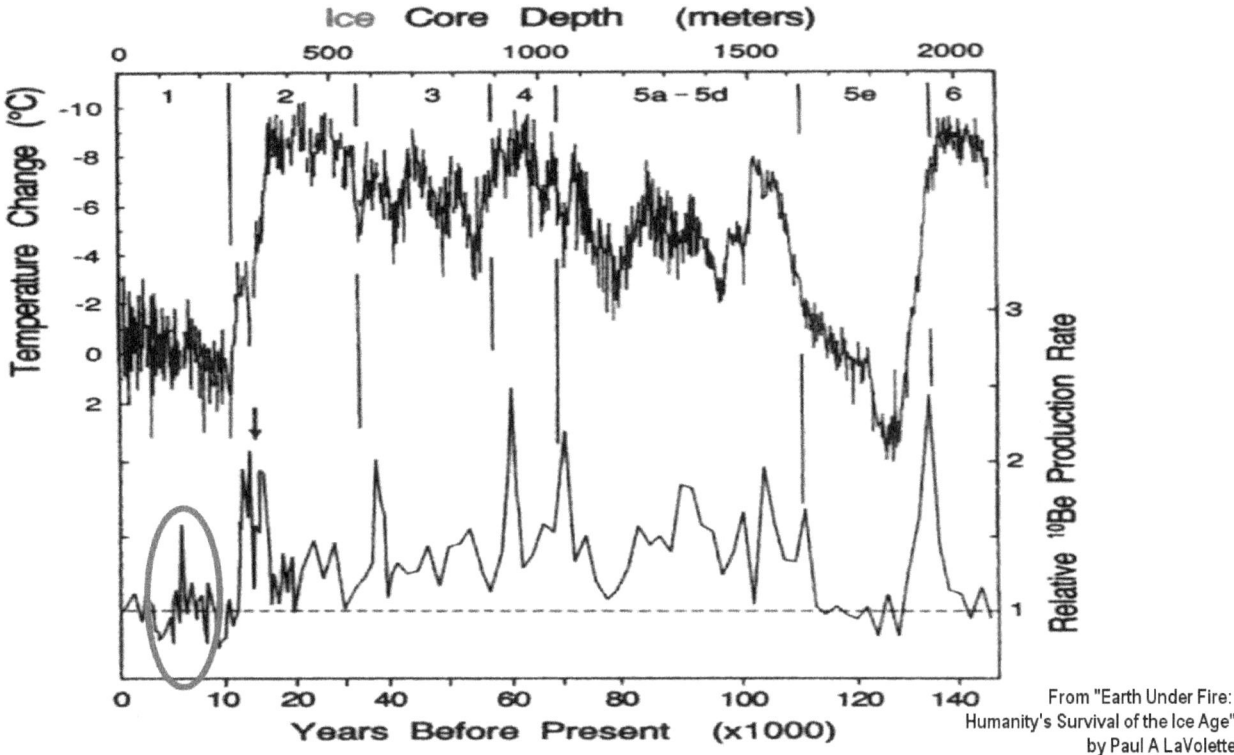

The first thing that stands out, is the spike in Beryllium production at roughly 5000 years before the present. What does this spike tell us?

Greenland ice core at 4770 years before the present

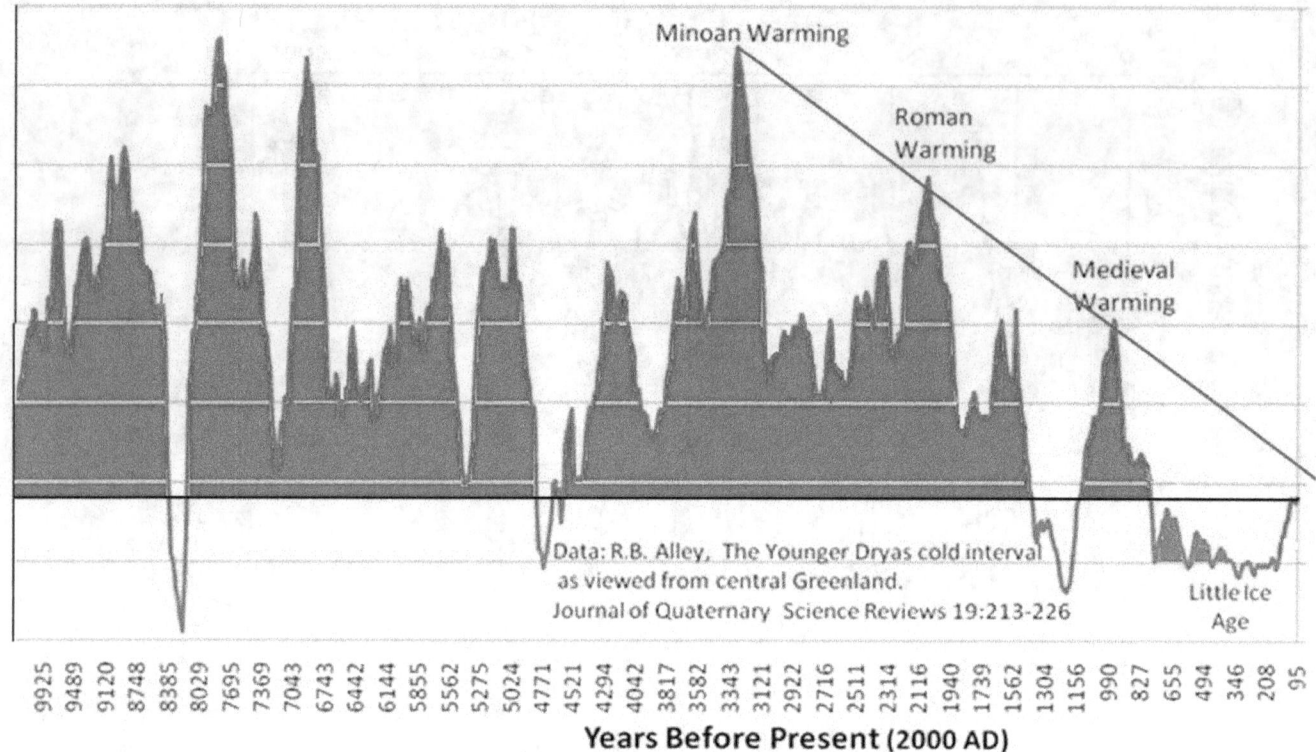

If we look at the Greenland ice core climate record, at the temperature value at 4770 years before the present, we see a big low-temperature climate indicated. This temperature drop took us as low in the temperature record, if not lower, than the little Ice Age had taken us in later years.

The coincidence of the Beryllium spike

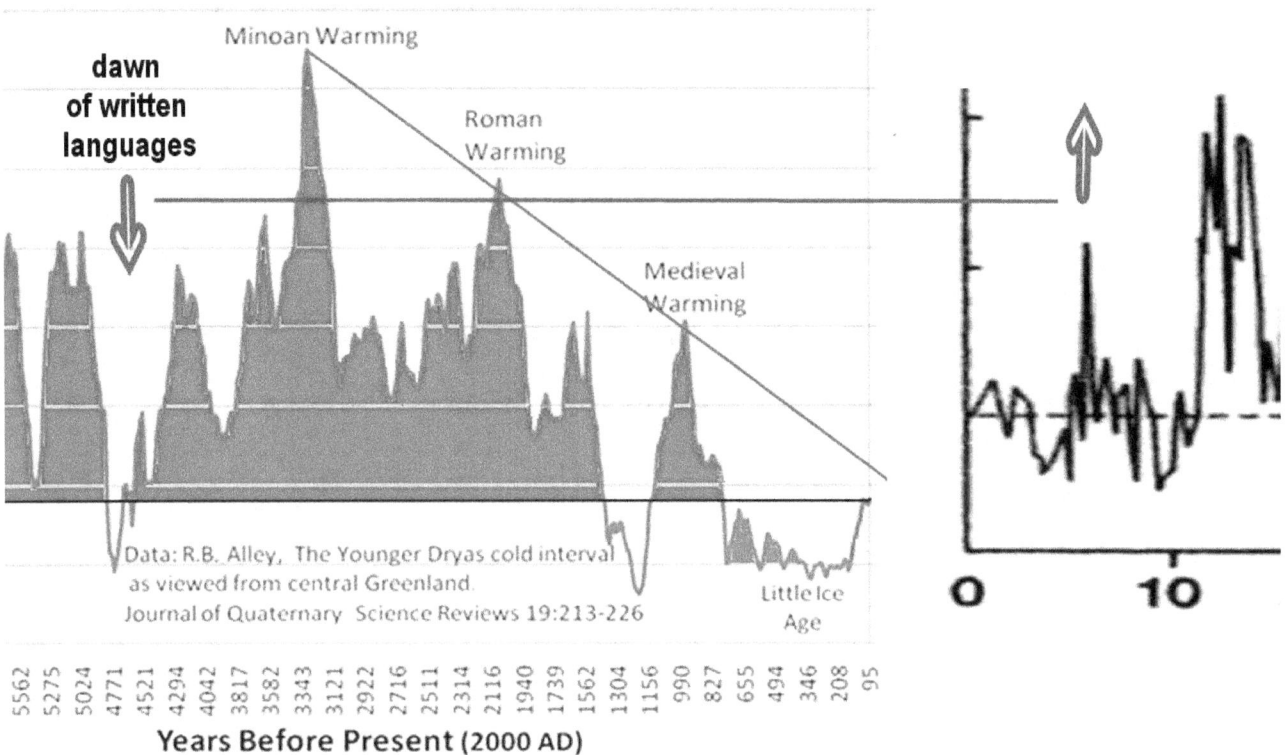

The coincidence of the Beryllium spike for this timeframe, shown expanded here, tells us that this period of deep cold climate was caused by the Sun. When the Sun weakens, its surrounding plasma sphere is likewise weaker, which enables more cosmic-ray flux to escape from the Sun and affect the Earth.

We see the result here in the form of an increased Beryllium production in the atmosphere. That's what the spike represents.

Colder climate is the result of a weaker Sun

ISS-34 - Stratocumulus clouds

We also see the result of cosmic-ray ray showers affecting the Earth in the form of increased cloudiness.Increased cloudiness makes the Earth colder, because the white top of clouds reflect a greater portion of the sunlight back into space, which thereby becomes lost to us. The resulting colder climate is thereby the result of a weaker Sun.

Cosmic-ray flux ionizes water vapor

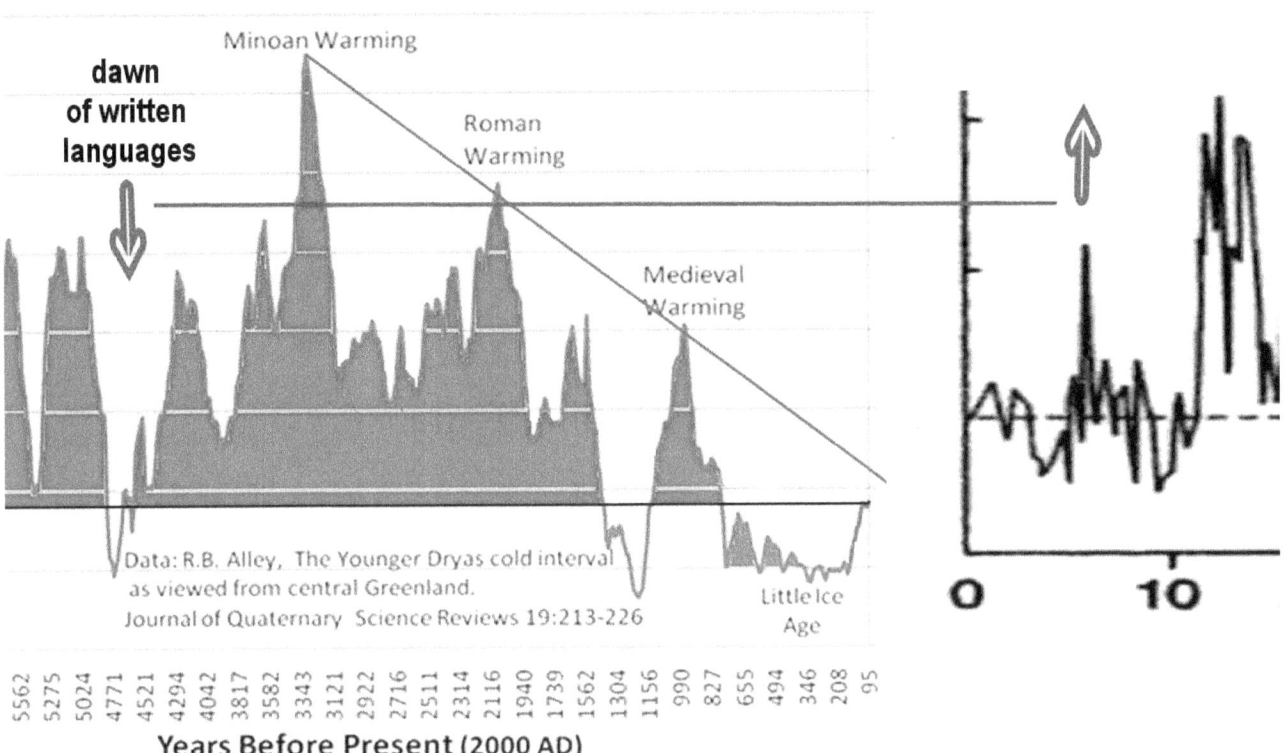

Cosmic-ray flux ionizes the water vapor in the atmosphere, which leads to increased cloudiness.

The spike in Beryllium, for this time, tells us that the Sun was extremely weak at this time, and was emitting high rates of cosmic rays, which produced high rates of Beryllium-10.

Some type of anomaly had evidently affected the plasma supply stream of the Sun for a few decades; and consequently, as it must, the climate had responded by getting colder.

This all by itself tells us that the Sun is the indisputable master of the climate on Earth. However, the Sun is not its own master. This means that the Beryllium production serves us as a measurable indicator of how weak the solar system as a whole had become at a specific time, and the Sun with it.

➤ The start of the current interglacial

The start of the current interglacial

The start of the current interglacial

Spikes between 11,000 and 15,000 years be fore the presen

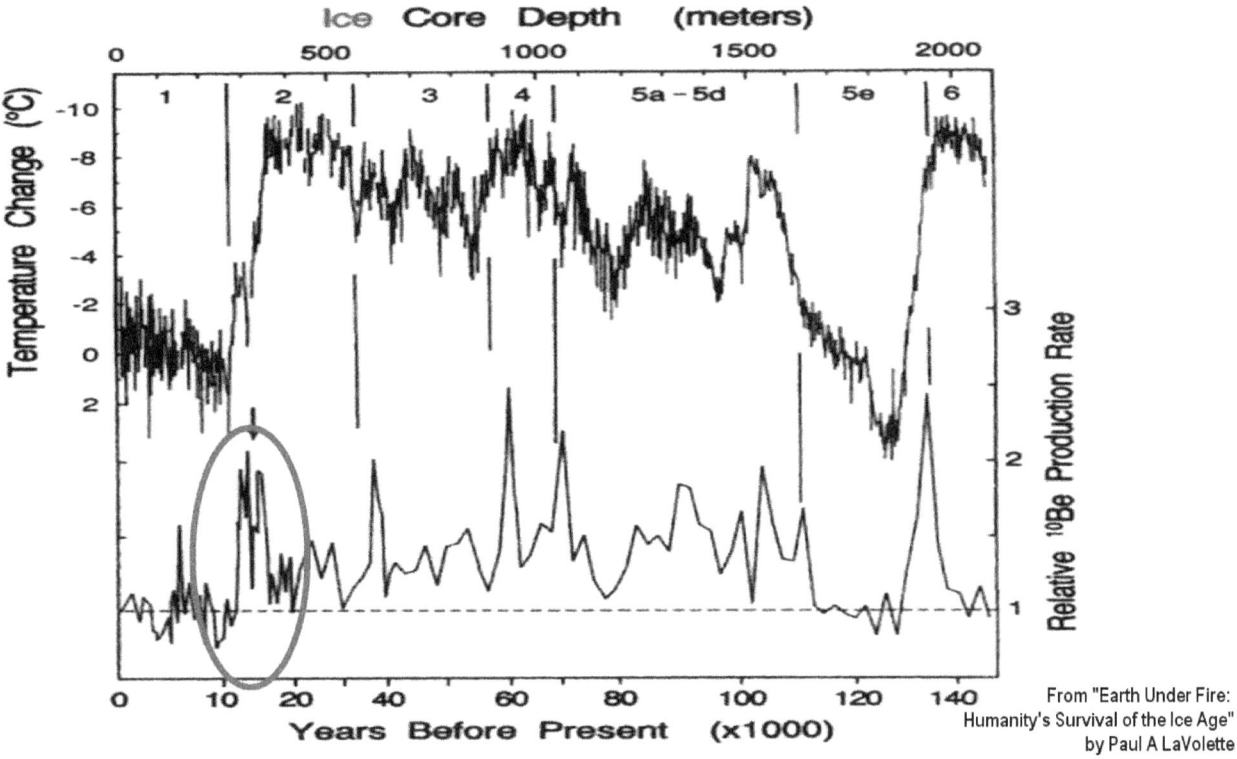

From "Earth Under Fire: Humanity's Survival of the Ice Age" by Paul A LaVolette

Next, let's look at the two consecutive spikes between 11,000 and 15,000 years be fore the present.

These spikes shouldn't be there. We are still in glacial climate. The spikes seem to indicate that the interstellar plasma stream was getting stronger again, strong enough to get the Sun ramped up again, to awake it from its long glacial low-powered hibernation.

The two spikes are evidently not the result of an extraordinarily weak Sun, with a weak plasma shield around it. Instead they represent the result of a weak Sun being ramped up. The weak Sun had almost no plasma shield around it during its glaciation time. When a hibernating Sun gets ramped up again, large volumes of solar cosmic-ray flux are generated by the suddenly active Sun that has no active shield around it at the time, so that large volumes of cosmic-ray flux are able to escape.

The two spikes are telling us that the Sun was actually restarted twice. The first restart appears to have failed.

The first startup, 15,000 years ago failed

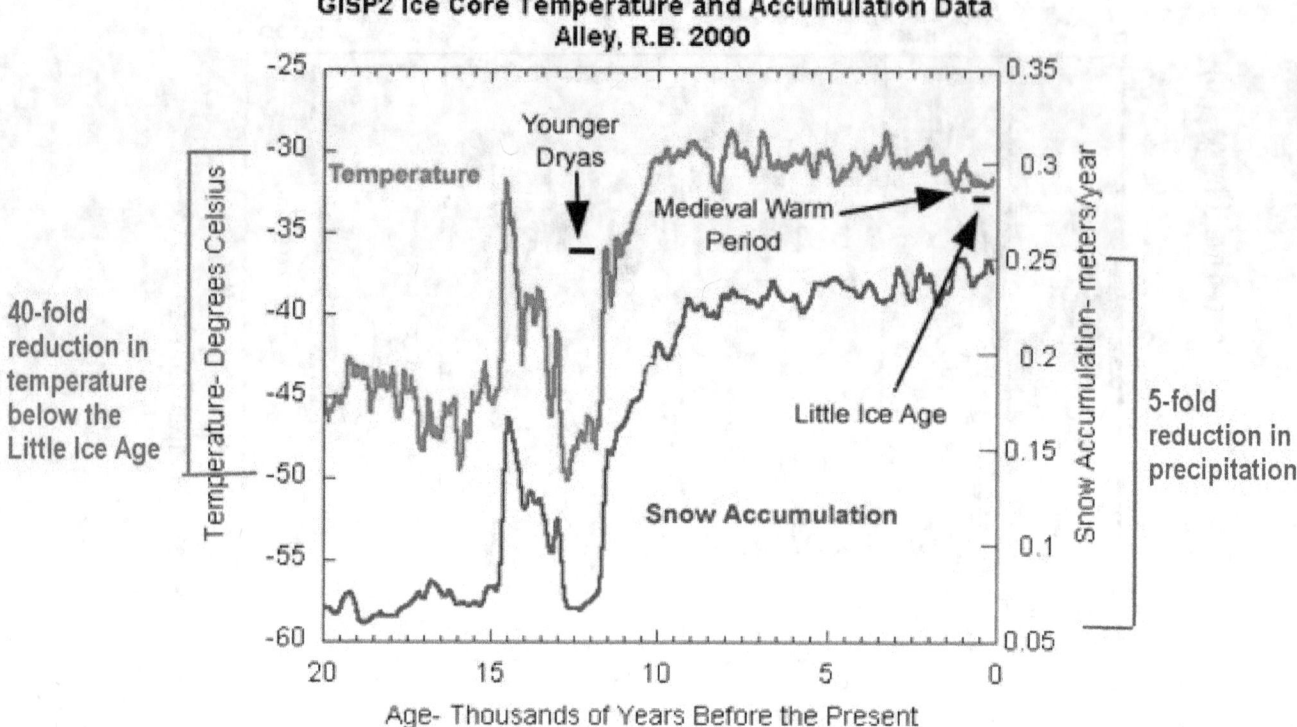

The plasma inflow was evidently not dense enough for the first startup, 15,000 years ago, to sustain full-power solar operation. The startup failed and glaciation conditions resumed for another thousand years till the plasma density was sufficient to form primer fields again that focus plasma unto the Sun, with enough plasma-density inflowing this time for the startup to succeed.

The two-step process to get the Sun ramped

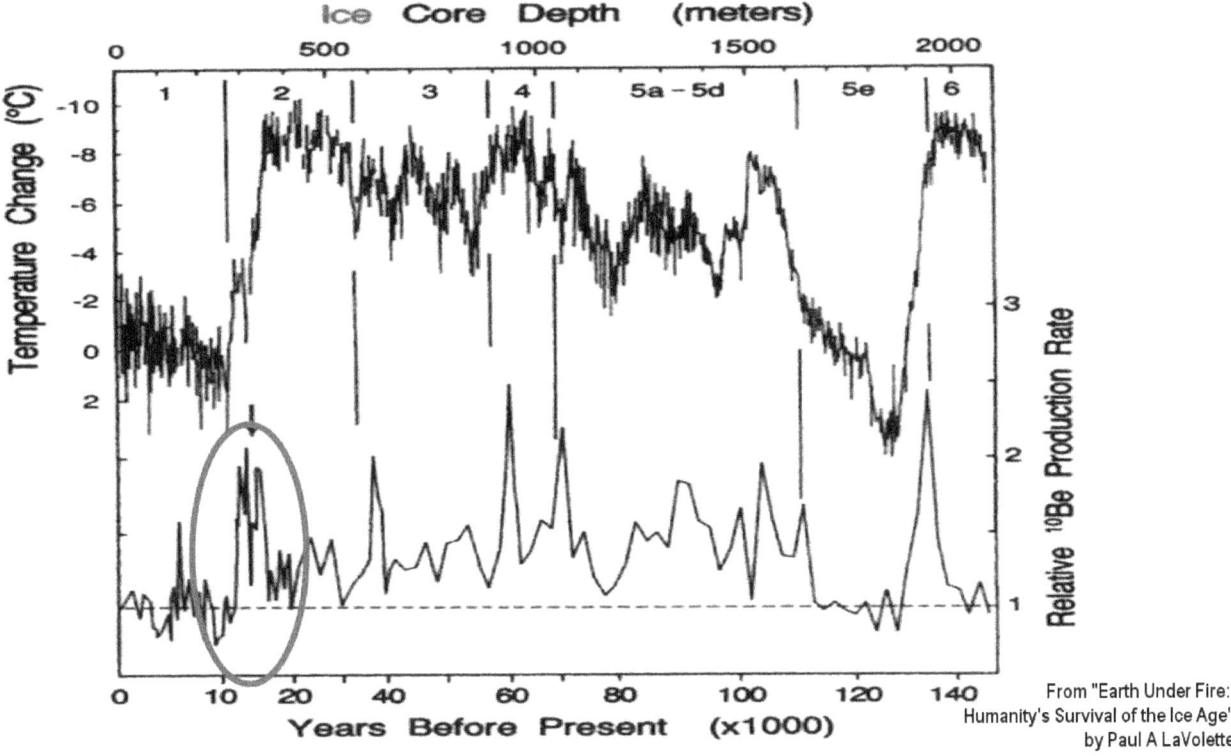

We see the two consecutive Beryllium pulses reflecting the two-step process to get the Sun ramped up to the current interglacial high-power mode.

After the startup succeeded, the Beryllium pulse ended, because by then, the fully-powered Sun had a dense plasma sphere around it, in which most of the solar cosmic-ray flux becomes trapped.

In the mechanistic universe

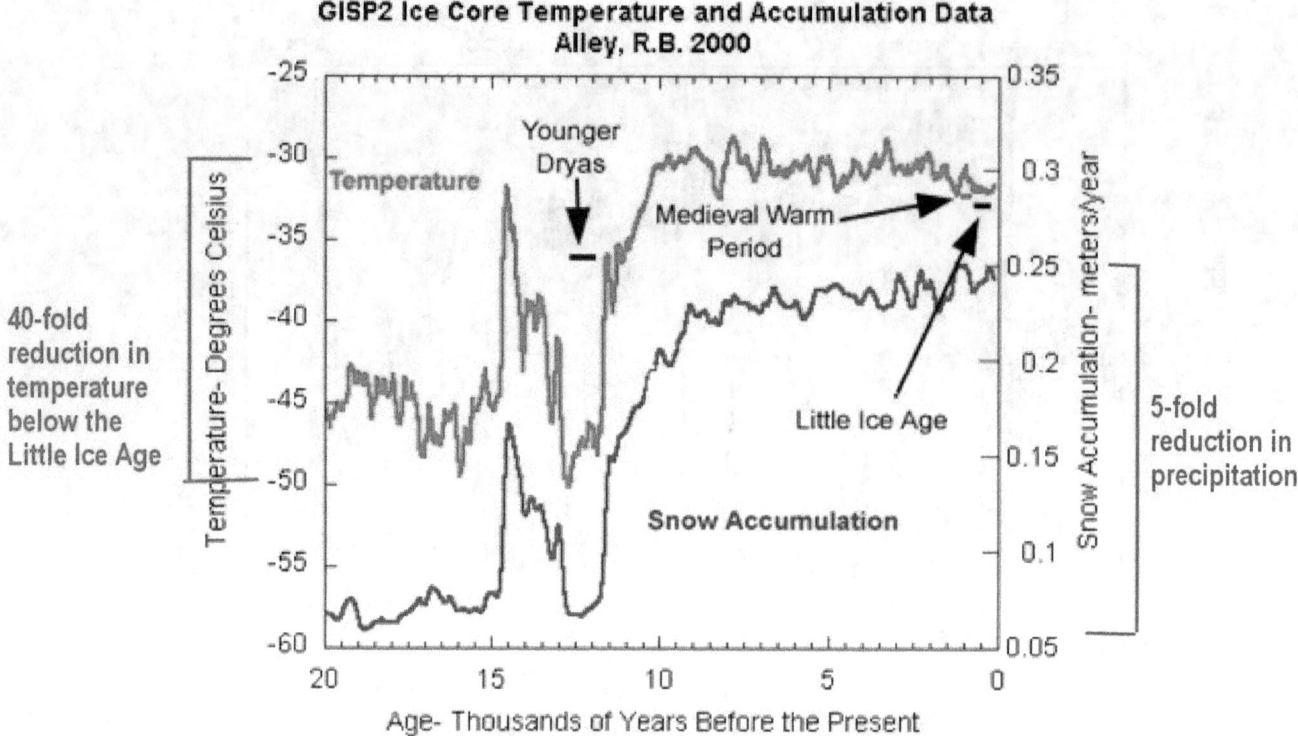

In the mechanistic universe, in which the Sun is deemed to be invariable and not a climate factor, the end of the last Ice Age is placed at 15,000 years ago. It is reasoned that the end of the Ice Age brought so much melt water into the oceans that it changed ocean currents, which is deemed to have brought the Ice Age back. This is a long-believed theory.

The two spikes in the Beryllium record

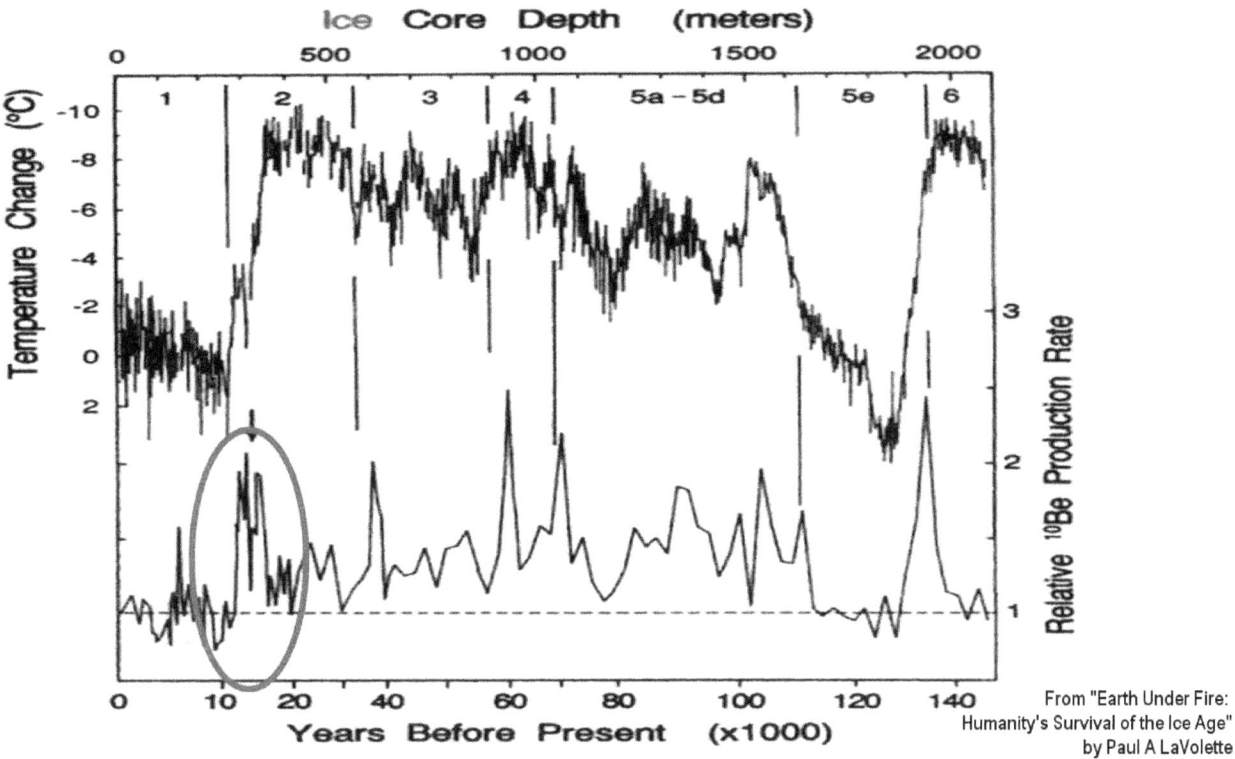

From "Earth Under Fire: Humanity's Survival of the Ice Age" by Paul A LaVolette

However, the two spikes in the Beryllium record now prove this theory to be invalid.

> **The recovery of the Sun 130,000 years ago**

**The recovery of the Sun
from the previous Ice Age
130,000 years ago**

The recovery of the Sun from the previous Ice Age 130,000 years ago.

The Beryllium spike 130,000 years ago

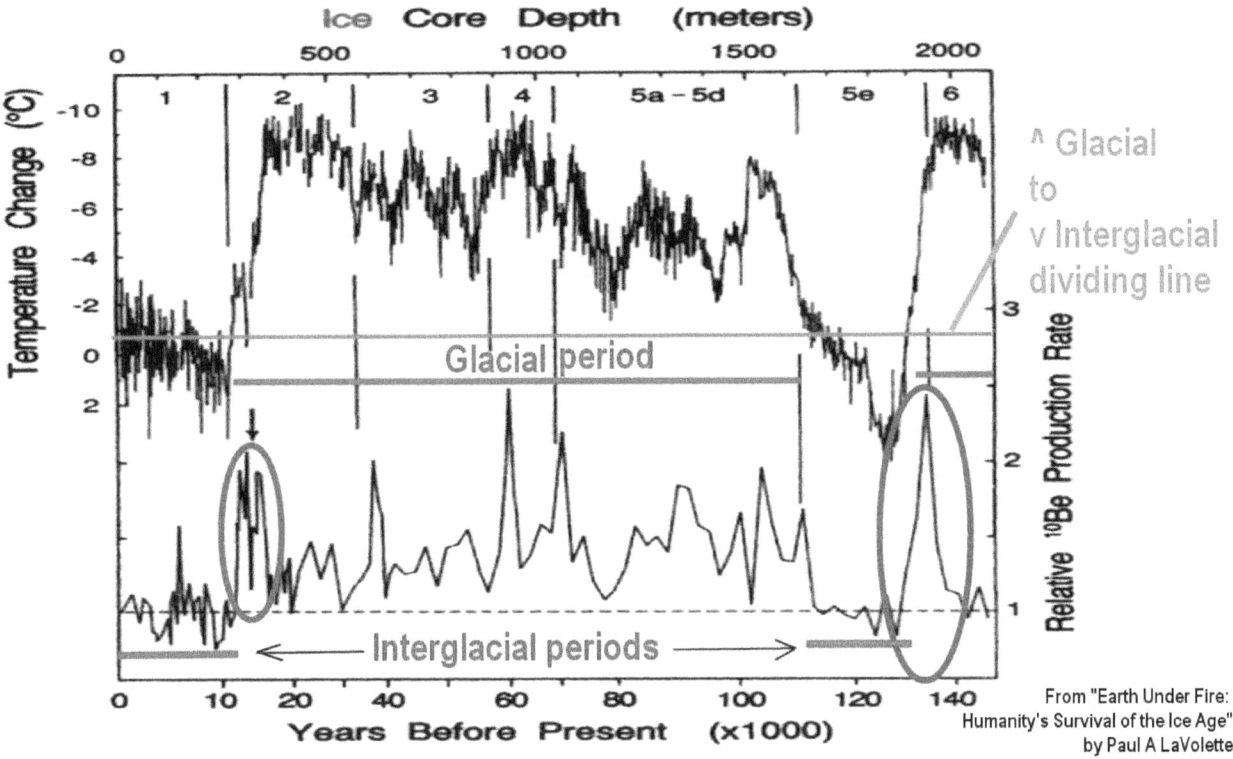

The Beryllium spike that occurred 130,000 years ago, evidently resulted from the plasma input-streams to the Sun recovering in density to interglacial levels. The plasma density pulse at 130,000 years ago was evidently strong enough to up-ramp the Sun all the way through to its continuous operation, without a failure on the way. The end of the pulse ushered in the previous interglacial period.

The large spikes that we find between the end of the glacial periods and the start of the interglacial periods, tell us that the Ice Ages were caused by the Sun hibernating during the glacial times, and the Sun's reawakening with strengthened input plasma streams that generate high-volumes of solar cosmic-ray flux on the Sun as the Sun rebuilds itself to its interglacial high-power state.

This means that the two little spikes between 11,000 and 15,000 years ago, and the big one 130,000 years ago tell us the same story, and they tell us that the time has come to rethink our perception of the universe, and consider the plasma dynamics as the driver of our Sun, and the Sun as being the climate driver on Earth, as a changing engine that affects us on Earth in a big way, including with the causing of the Ice Ages.

When the solar awakening is complete

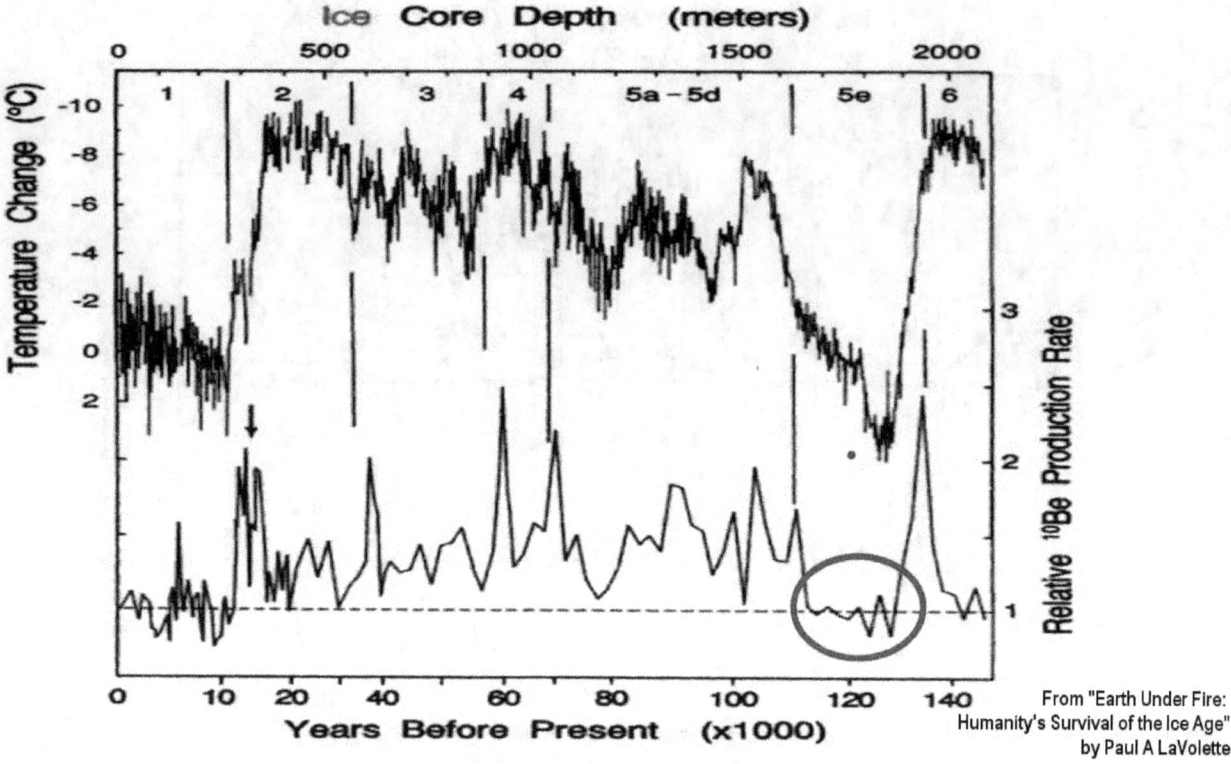

Also note that when the solar awakening is complete, the cosmic-ray flux returns to a low volume all the way through the interglacial period, as shown in the Beryllium record.

When the interglacial period ends

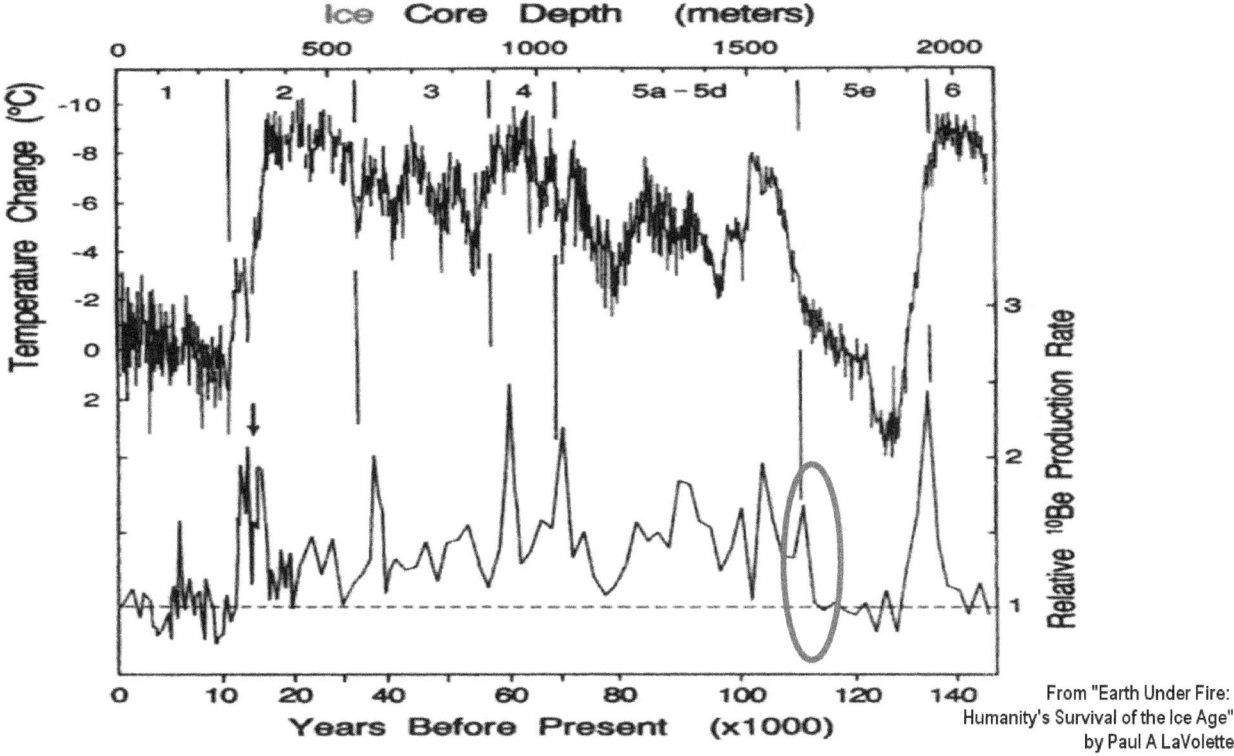

From "Earth Under Fire: Humanity's Survival of the Ice Age" by Paul A LaVolette

Then, when the interglacial period ends, when the primer fields collapse that focus plasma onto the Sun, the Sun reverts back to its low-power hibernation state that creates the glaciation period, which is a state of relatively high cosmic-ray flux for the lack of a shield against it.

> cultural effect of cosmic-ray flux

$$\frac{\text{The interglacial Sun versus the glacial Sun}}{\text{The cultural effect of cosmic-ray flux}}$$

The interglacial Sun versus the glacial Sun. - The cultural effect of cosmic-ray flux.

High Beryllium level during the glaciation period

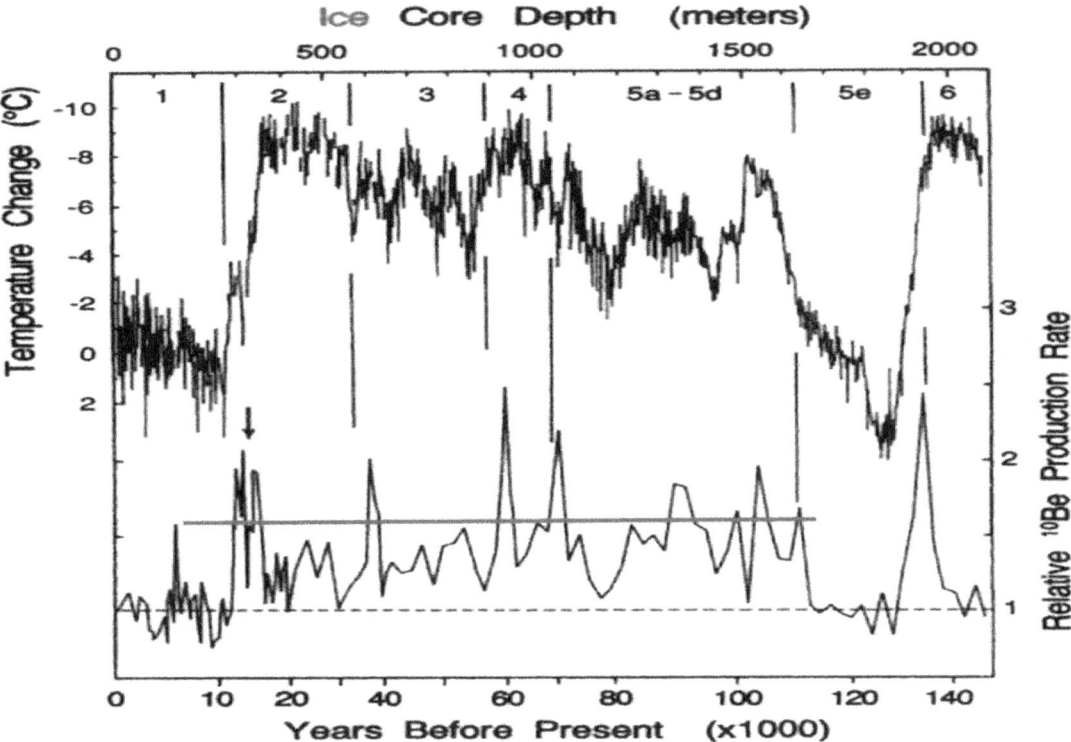

It is interesting to note that for the entire glaciation period, the plotted Beryllium record is substantially high all the way through the period. It should be noted too, that the high Beryllium level during the glaciation period was produced by cosmic-ray flux originating from the unshielded 'hibernating' Sun, which is a relatively cold Sun, potentially 70% colder than the present Sun. This means that during glacial times the cosmic-ray flux is not the only determining climate factor. The colder Sun is instead the major climate factor, while the cosmic-ray flux does have a minor effect on the Ice Age climate.

During interglacial times

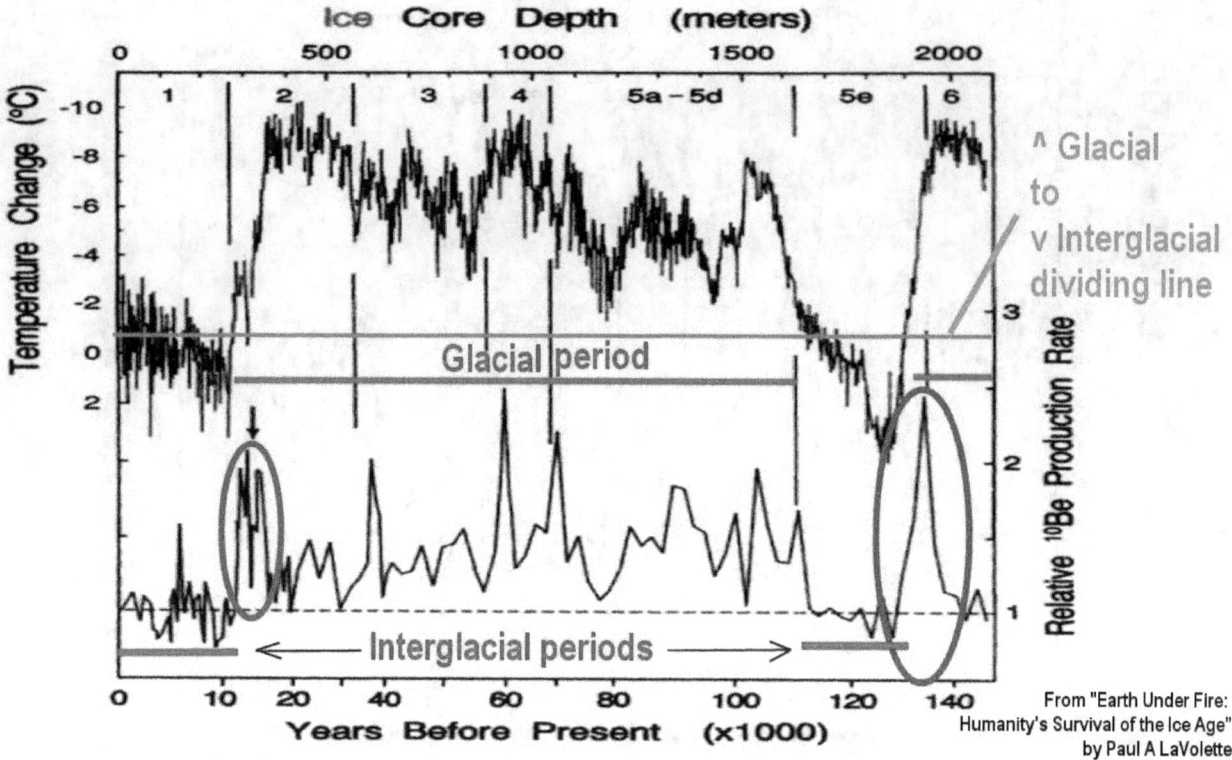

From "Earth Under Fire: Humanity's Survival of the Ice Age" by Paul A LaVolette

Inversely, during interglacial times, the cosmic-ray flux is a strong climate factor, because its ionizing effect affects cloudiness in a big way. Cloudiness is strong during interglacial times. The warm oceans under the hot interglacial Sun evaporate large volumes of water vapor.

During the Ice Age glaciation period, with 80% less moisture in the air, cloudiness is a lesser factor, though it is a factor in the form of ice fog. The determining factor during the Ice Age glaciation period is always the low surface temperature of the hibernating Sun, which is also changing during glacial times.

Beryllium level during the glaciation period

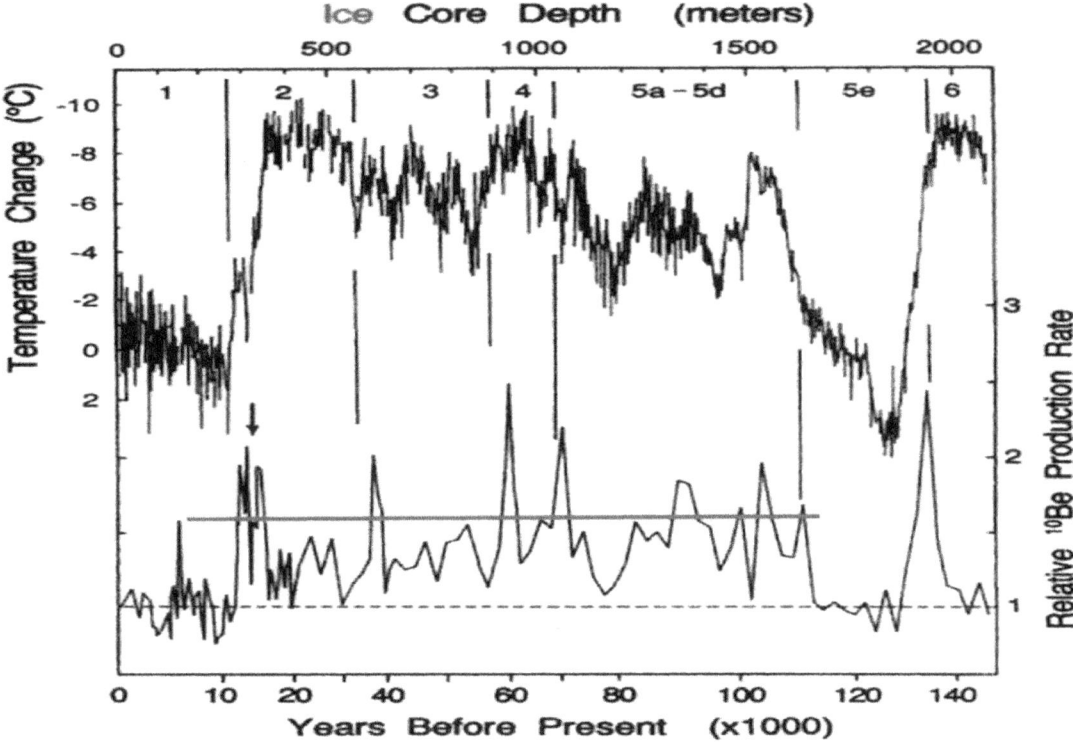

It should be noted too that the average Beryllium level during the glaciation period was roughly at the same level as that of the spike 4700 years ago. This is significant, though not for climate reasons.

The Beryllium spike 4700 years ago, is important

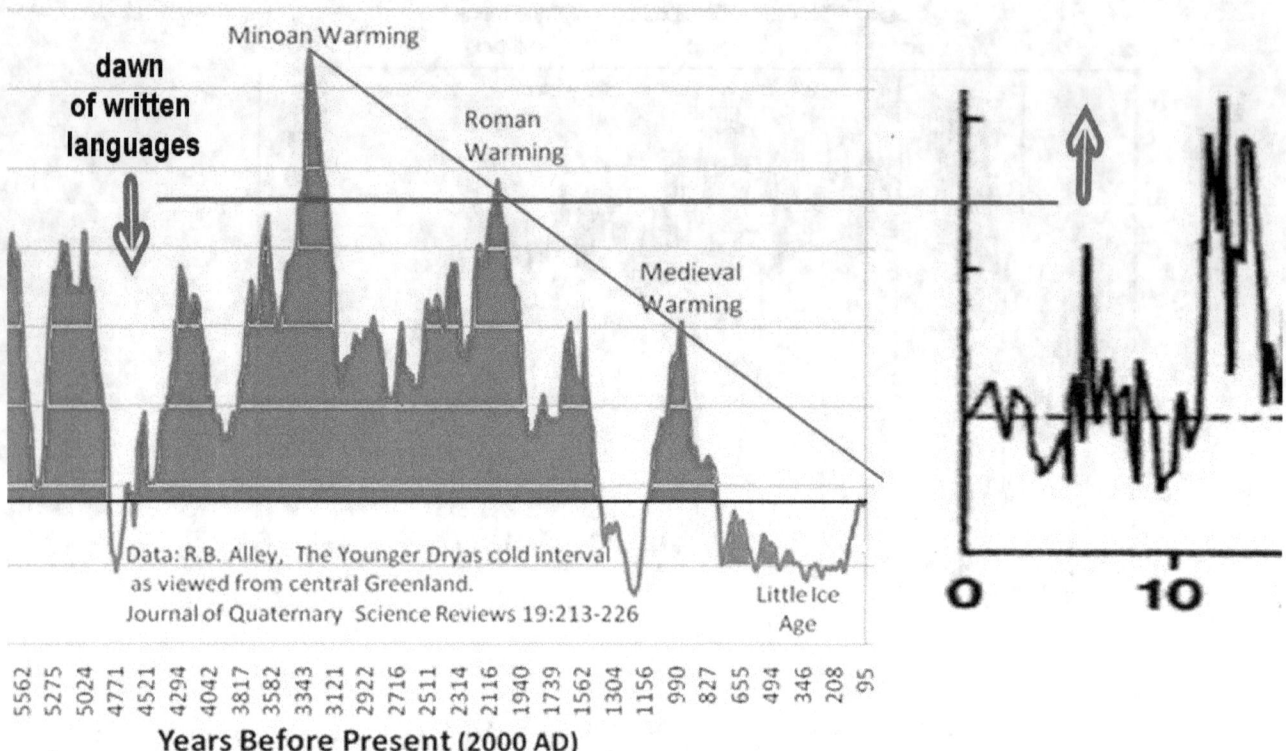

The high level of the Beryllium spike 4700 years ago, is important, because, as I have previously indicated, the high rate of cosmic-ray flux that had produced that spike 4700 years ago, had also caused a phase shift in human history. We see such high Beryllium levels all the way through the last Ice Age, to varying degrees. And we will see it again through the next Ice Age.

The deep-cold period around 4700 years ago

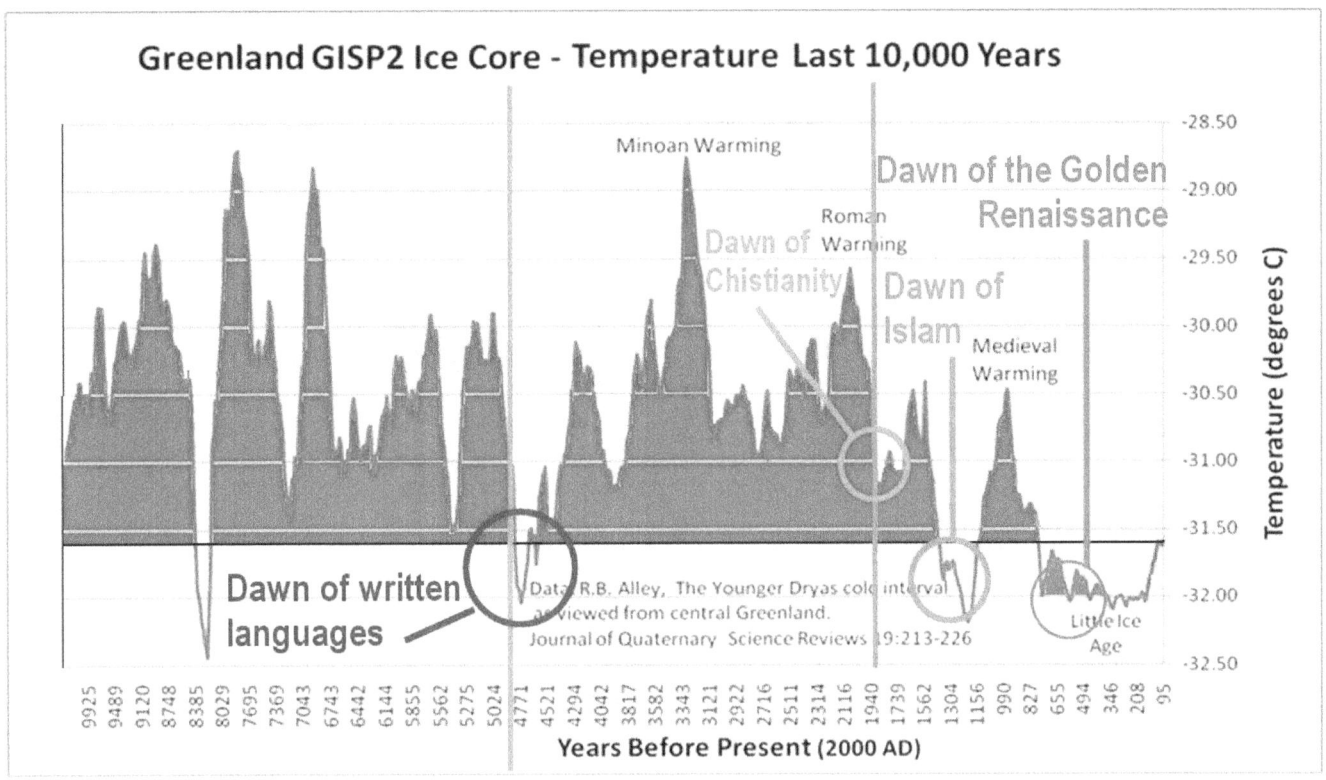

The deep-cold period around 4700 years ago, which reflects the high-volume cosmic-ray background measured in Beryllium-10, was the historic timeframe in which written languages were invented and came into general use. A high-volume cosmic-ray background appears to have this effect. This type of cultural uplift is what we will likely experience all the way through the next Ice Age.

A minuscule trail of induced electric currents

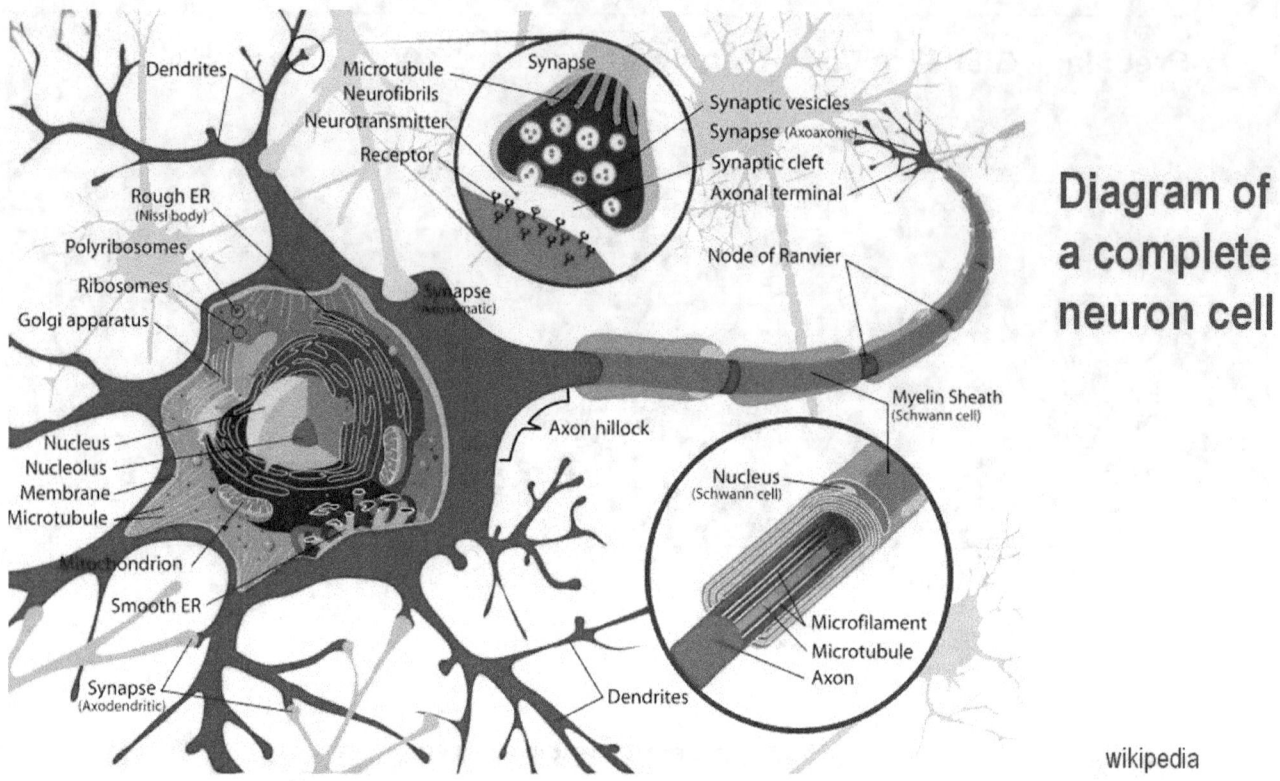

Diagram of a complete neuron cell

wikipedia

Cosmic-ray particles are fast moving plasma particles, primarily protons, which are 100,000 times smaller than the smallest atoms. They are so tiny, that they pass through a human body without colliding with anything. But as they path through us, their moving electric charge leaves behind a minuscule trail of induced electric currents that appear to be beneficial for human cognitive development.

All the great cultural achievements

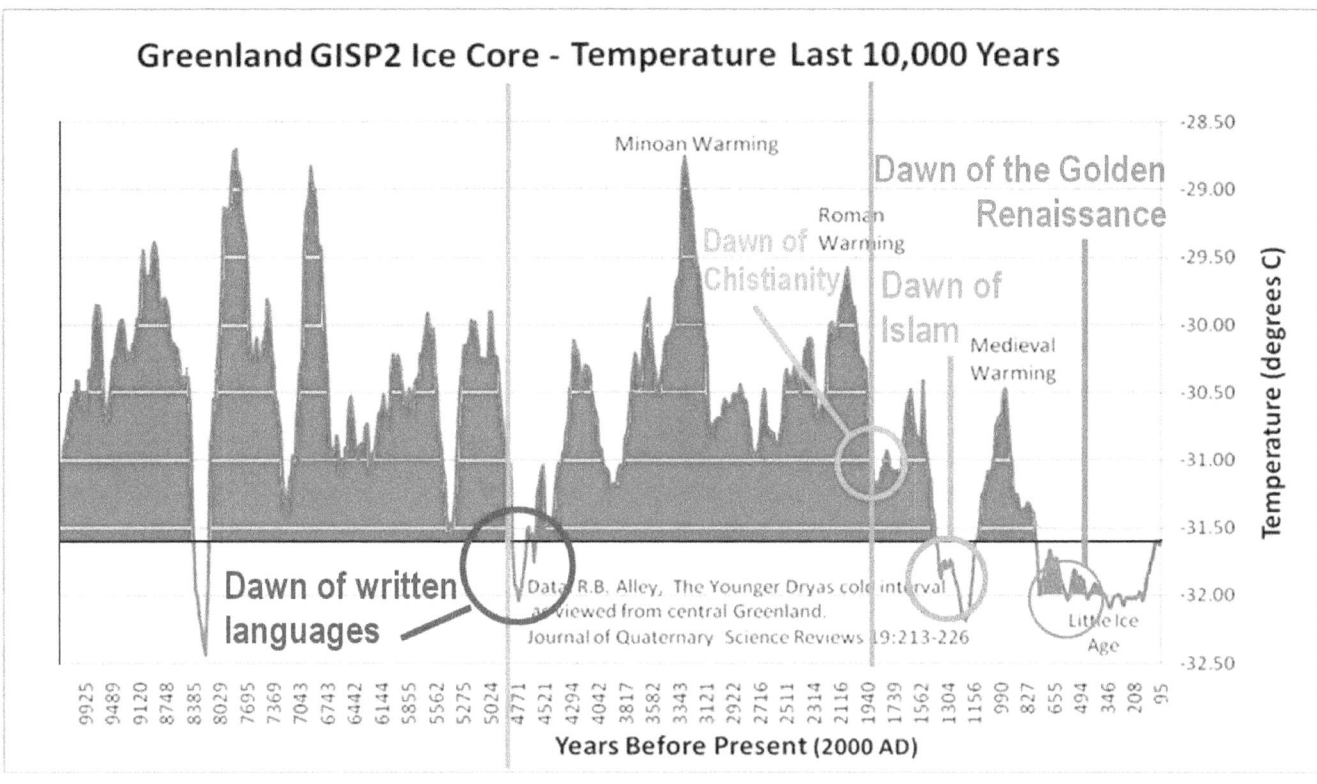

History tells us that all the great cultural achievements were wrought or started in the cold times that are times of high-volumes of cosmic-ray events.

It is not surprising in this context that the biggest breakthrough, the dawn of high-level written languages, which furnished the foundation for all the subsequent cultural advances, including that of science, technologies, and energy development, occurred at the time of that biggest spike in cosmic-ray flux 4700 years ago, right in the middle of the development of our high-level civilization. That spike may have enabled us to be what we are today.

A similar high-level cosmic-ray background

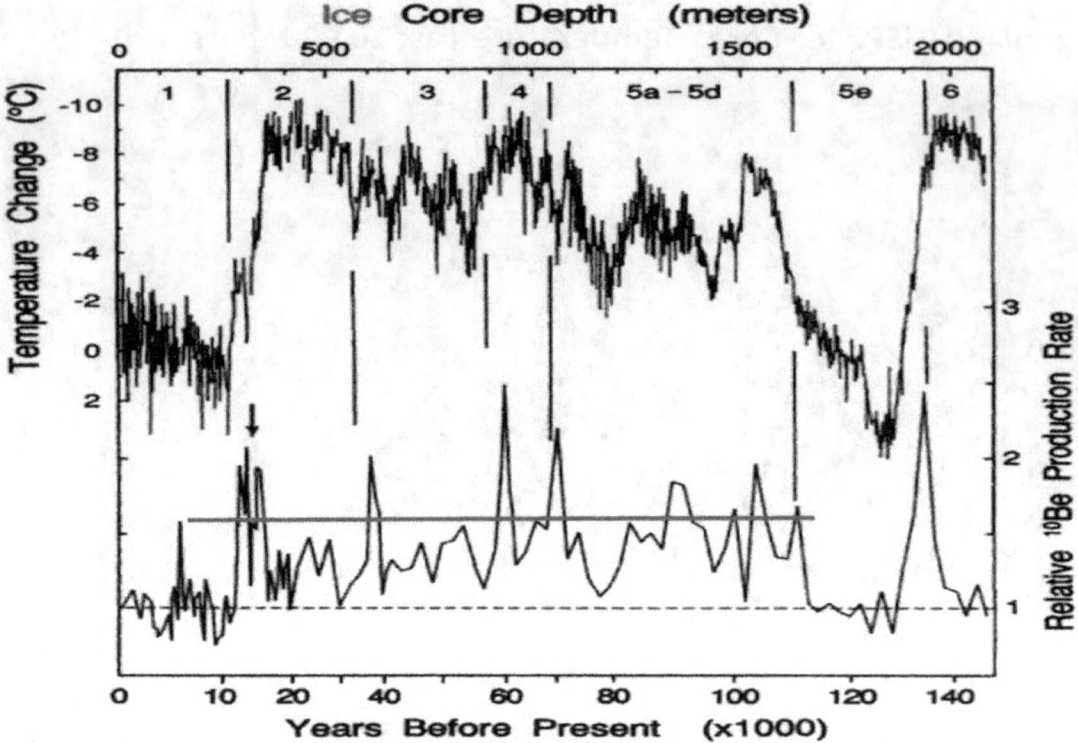

It is interesting to note here that the big Beryllium spike at 4700 years ago, is roughly equal to the average Beryllium level all the way through the last glaciation period. This means that we have some revolutionary cognitive, cultural, scientific, even technological and spiritual developments to look forward to as we enter the next glaciation period that promises a similar high-level cosmic-ray background.

We may be closer to this happening than we yet realize.

➢ The Earth's magnetic polarity reversal

The famous 41ka Beryllium spike
───────────────────────────────
and the Earth's magnetic polarity reversal

The famous 41ka Beryllium spike and the Earth's magnetic polarity reversal.

The famous Beryllium pulse 41,000-years ago

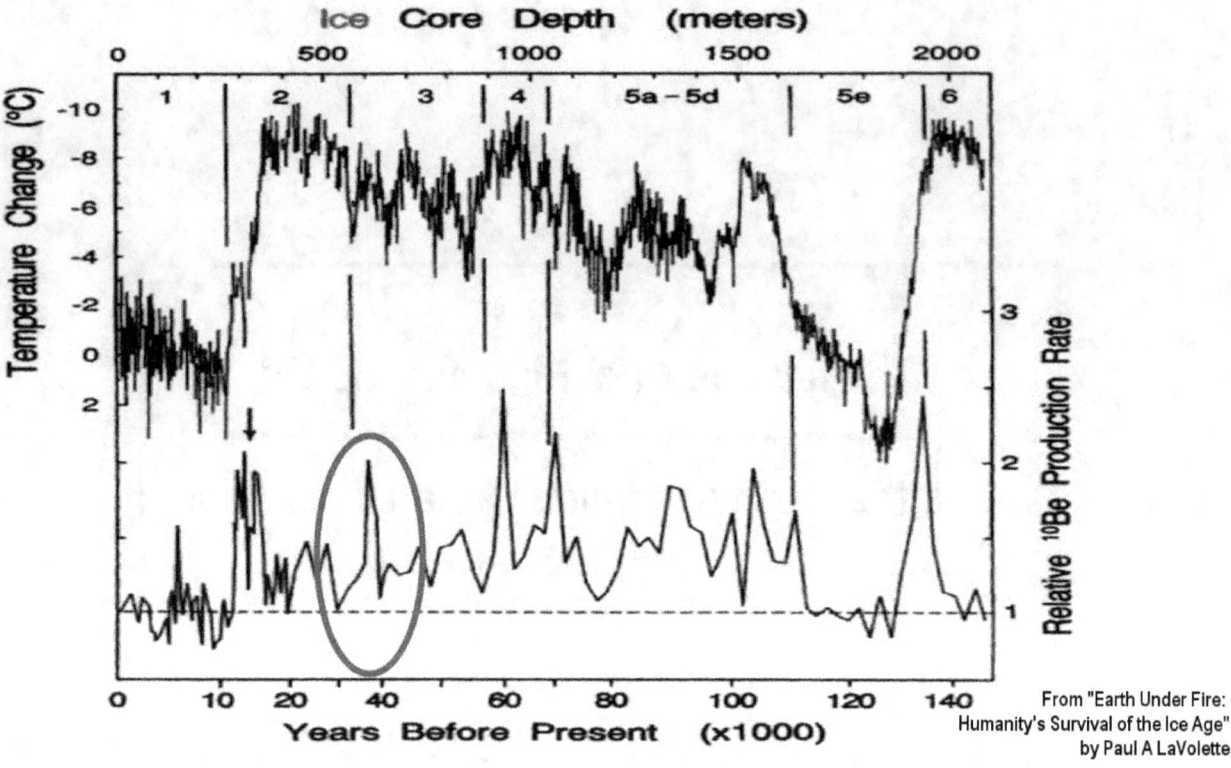

From "Earth Under Fire: Humanity's Survival of the Ice Age" by Paul A LaVolette

The famous Beryllium pulse 41,000-years ago spiked to the double mark. It may have started out as an awakening of the Sun, right in the middle of the Ice Age, which eventually failed.

It has been discovered that 41,000 years ago a magnetic pole reversal on the Earth had occurred, for a brief period of 250 years. The cause remains a puzzle. The polarity reversal would likely have resulted during the transition of the Sun from its low-powered glacial state to its high-powered interglacial state, which might even have briefly succeeded.

The Beryllium pulse at the 41,000-year mark would most definitely have been caused by strongly increased cosmic-ray flux flowing from the Sun, instead of it being galactic in origin. It is not possible for galactic, or extragalactic, cosmic-ray showers to flip the magnetic polarity of the Earth. But it is possible for this to happen by the solar primer fields starting up to some degree, out of balance and oriented against the Earth's magnetic polarity. The false startup process eventually failed and the Earth polarity reverted back to its normal polarity.

The magnetic polarity of the Earth

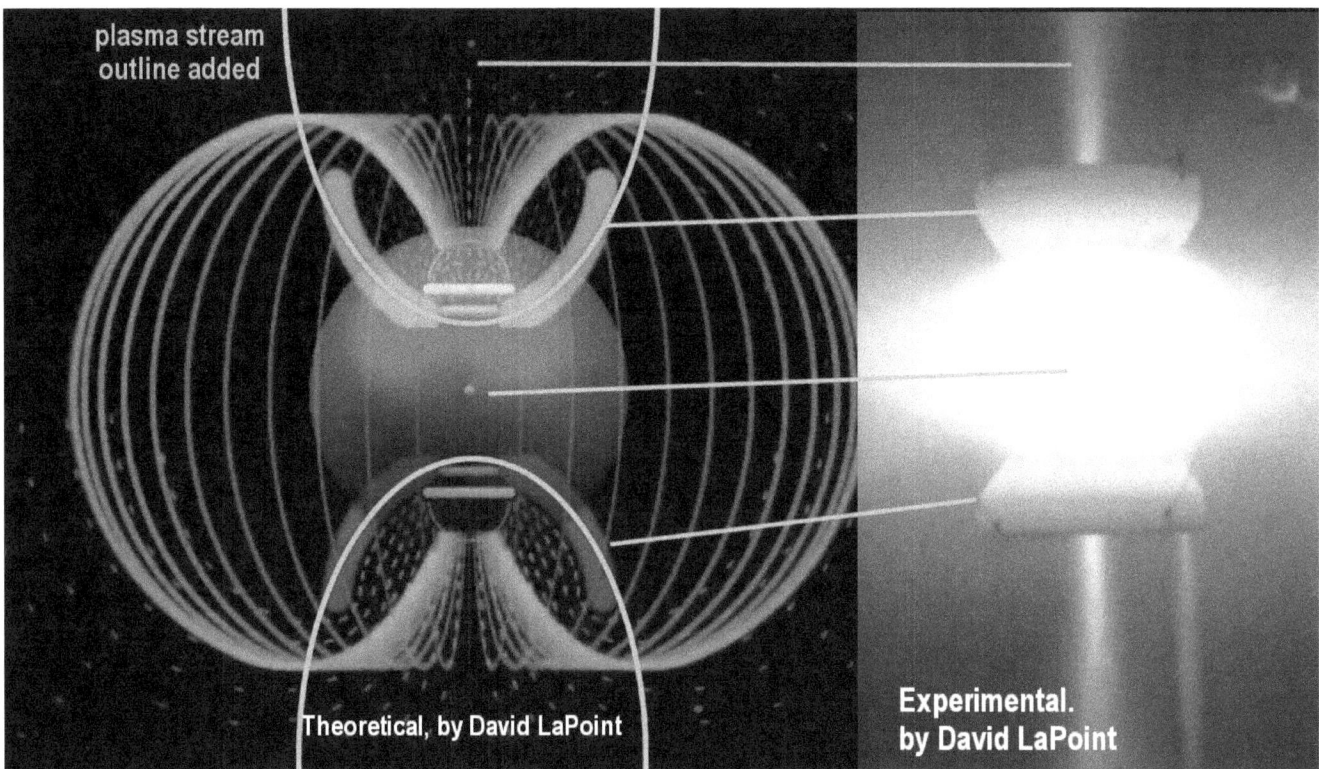

The Sun itself undergoes magnetic polarity reversals regularly between successive sunspot cycles. This is theorized to be the result of the resonance effect between the inner primer fields. The result appears to be too localized to affect nothing else but the Sun.

The magnetic polarity of the Earth, and the solar system as a whole, appears to be created by the dominance of one of two opposing gigantic magnetic structures that have the Sun at their center, and the planets located between them. A major imbalance can change the magnetic polarity of the Earth and possibly of the entire solar system. During the magnetic reversal 41,000 years ago, the Earth's magnetic field was extremely weak, at a mere 5% of today's level. The weak magnetic field would have left the Earth more exposed to solar cosmic-ray flux, which in this case might have been a major factor for the large Beryllium spike.

- The Dangaard Oeschger oscillations

> The Dangaard Oeschger oscillations

The Dangaard Oeschger oscillations

The researchers Dansgaard and Oeschger

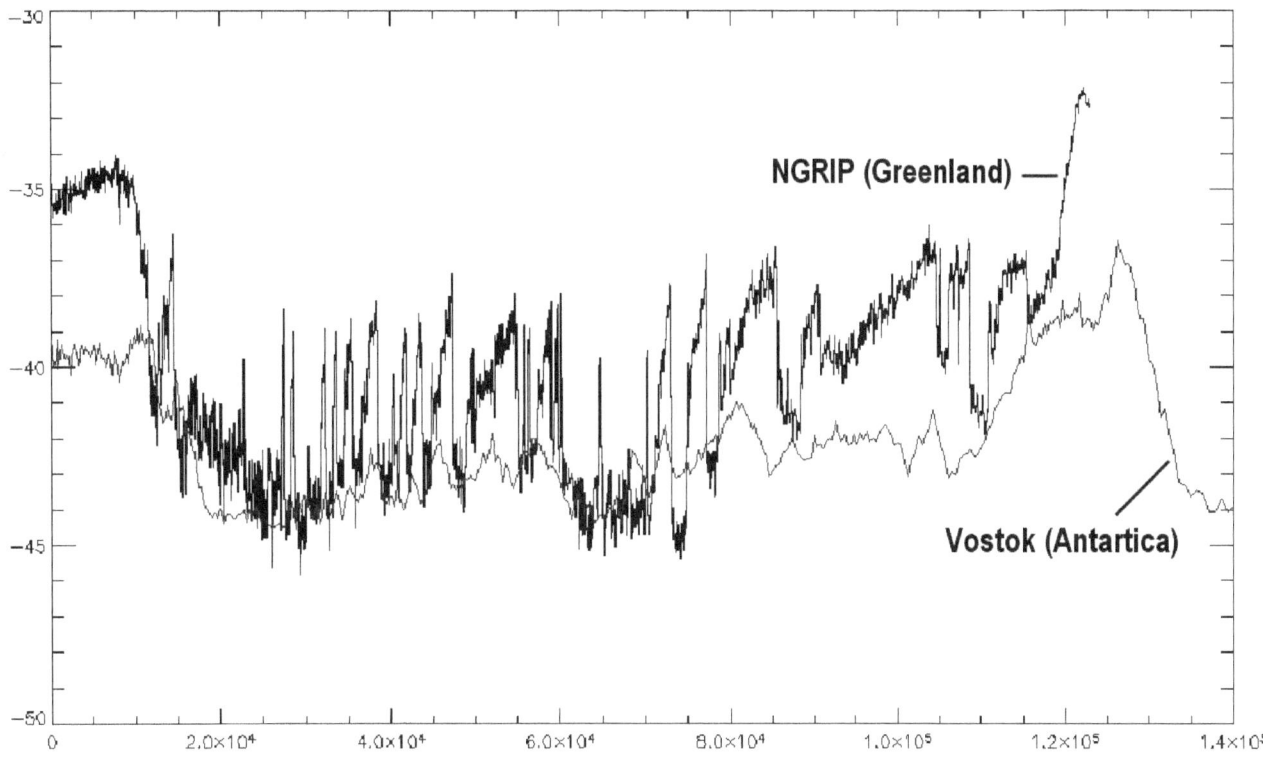

The researchers Dansgaard and Oeschger had sifted through the Greenland ice core climate records and had encountered large climate oscillations with sharp transitions from deep glacial conditions to near interglacial condition. They detected over 20 fast up-ramping-transitions, followed by slow decay times, typically spaced in intervals of 1470 years. No magnetic polarity reversals happened during these times. Nor am I aware of any proof in Beryllium measurements that these large oscillations were caused by the Sun, though it is highly unlikely that any other factor than the Sun is large enough to have caused the gigantic oscillations. It is more likely that the fast transitions, the fine details, including the slow down ramping had become lost by the low resolution of the ice in Antarctica.

On Antarctica, were the 95,000 years of glacial history is compressed into 1300 meters of Ice, or 250 months per foot of ice, the fine details tend to become washed out. For this lack of resolution the Dansgaard Oeschger oscillations are not detected in Antarctica, which is largely an ice dessert with minuscule precipitation, but they are detected in the Greenland ice in two different ice cores from widely separated drilling sites. Some day the Greenland ice cores may also be examined for their Beryllium-10 concentrations.

Temperature fluctuations in the Greenland ice cores

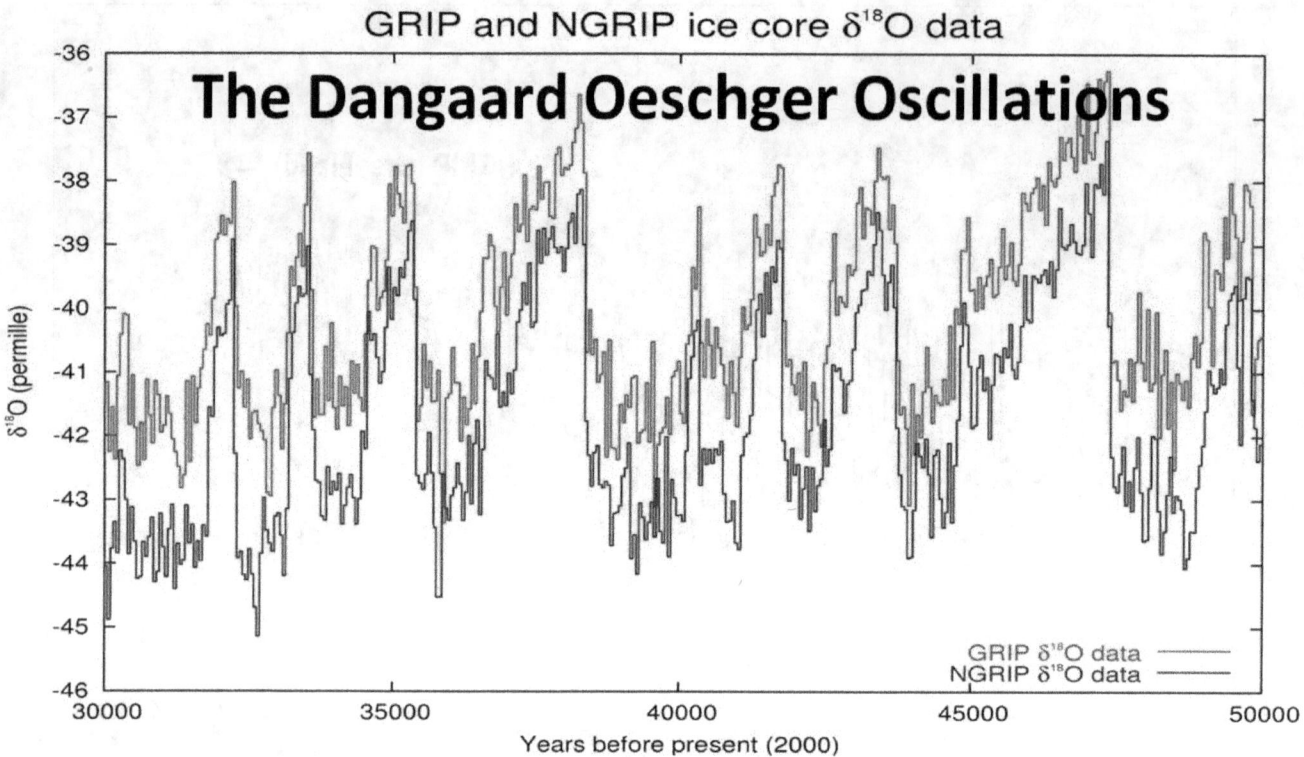

The temperature fluctuations that have been detected in the Greenland ice cores appear to be far too large and too rapid, for them not to have been caused by the Sun ramping up in some way in solar activity. If so, these solar effects would have caused correspondingly measurable proxy effects, which hopefully may be measured some day.

➢ The solar versus the galactic

The solar versus galactic cosmic-ray background

The solar versus the galactic cosmic-ray background.

Not all Beryllium values are solar activity proxies

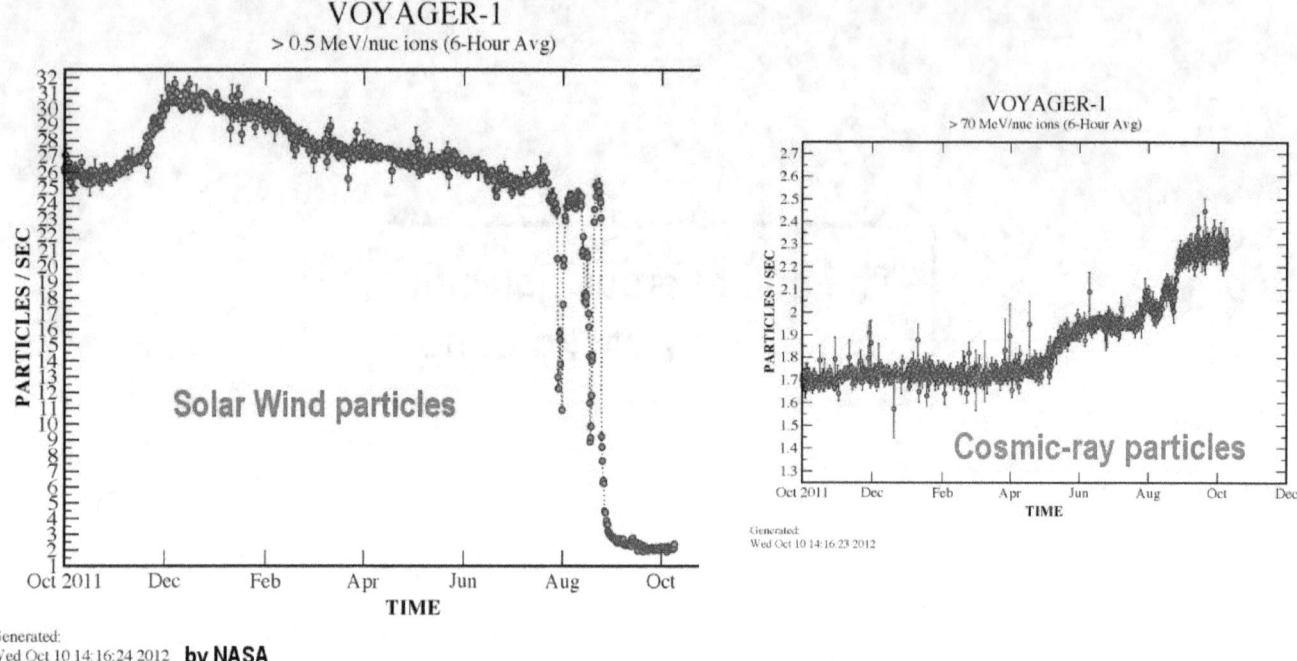

Not all the measured Beryllium values during the glaciation times, are solar activity proxies.

During glacial times, without the solar wind establishing a shield against galactic cosmic-ray flux, as we have it today in the form of the solar heliosphere, the galactic portion of the cosmic-ray flux would have been 35% greater. That's the amount that the heliosphere attenuates cosmic-ray flux during interglacial times according to measurements conducted by one of the Voyager spacecrafts that had penetrated the heliosphere.

This means that the measured Beryllium values, especially during the glacial period, do not exclusively stand as a proxy for solar activity, though a large portion of it evidently does. What the ratio is between galactic and solar cosmic-ray flux, during the glacial times, is left open to speculation.

The solar portion, however, appears to be the major contributing component. How enormously large this component can be, has recently demonstrated by different types of measurements of the solar cosmic-ray flux.

Sun is the major contributor to cosmic-ray flux

That the Sun is the major contributor to cosmic-ray flux in the solar system, has become strongly evident by the results of a recent series of laboratory experiments.

The researcher Simon Shnoll describes in his book "Cosmophysical Factors in Stocastic Processes" a series of chemical reaction experiments that he had continuously repeated in an identical manner over long periods of time. Theoretically the the results of these identical experiments should have been identical in every case, but they were not.

When the results were plotted

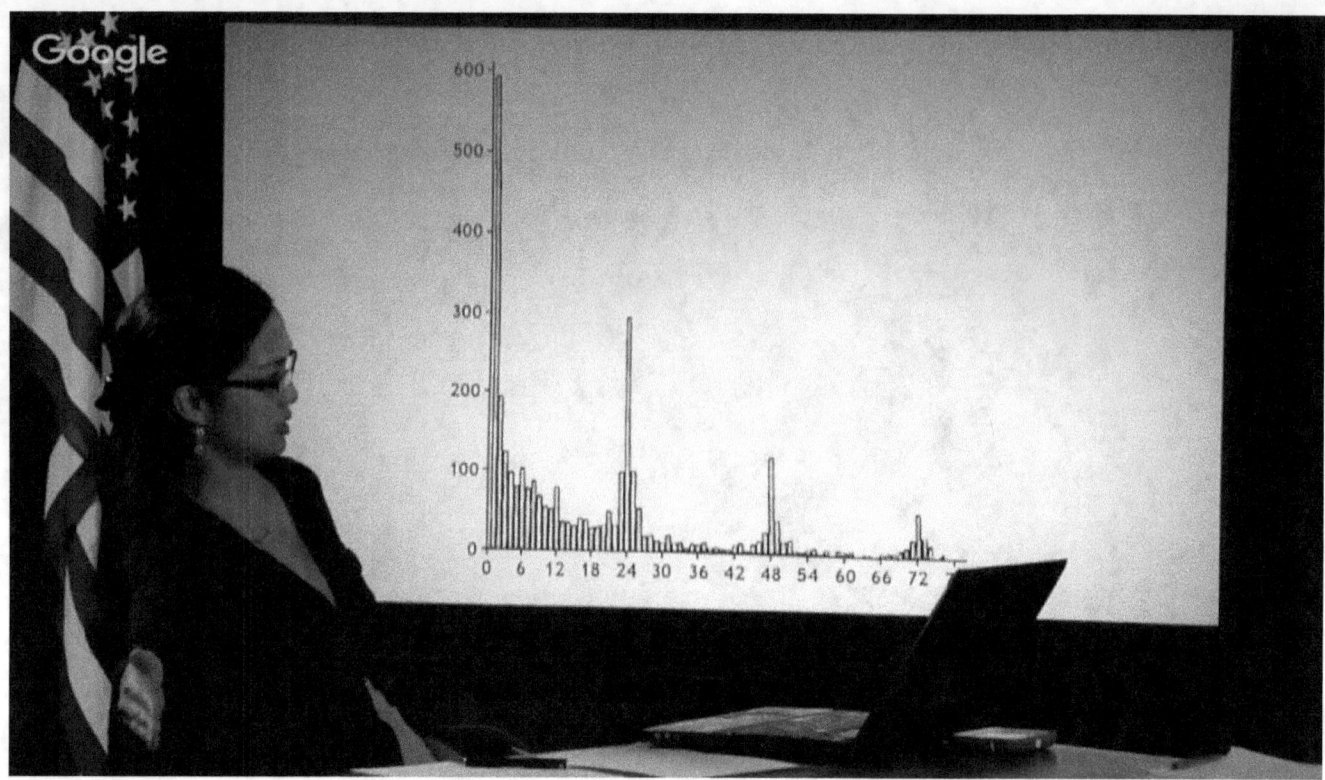

When the results were plotted, a large spike in the reactivity was detected which was repeated in 24-hours intervals, with the spikes getting progressively weaker. This too, was not expected.

It can be said, based on these results, that the 24-hours periodicity is related to the rotation of the Earth, so that cosmic-ray flux from a specific source could have caused these repetitive fluctuations. However, it cannot be said that the source was galactic in nature. The galactic flux doesn't change that quickly. The diminishing pulses, and the rate of diminishment, leaves the only one potential cause on the table, namely that the spiking was solar-caused.

When an Earth-oriented coronal hole opens up

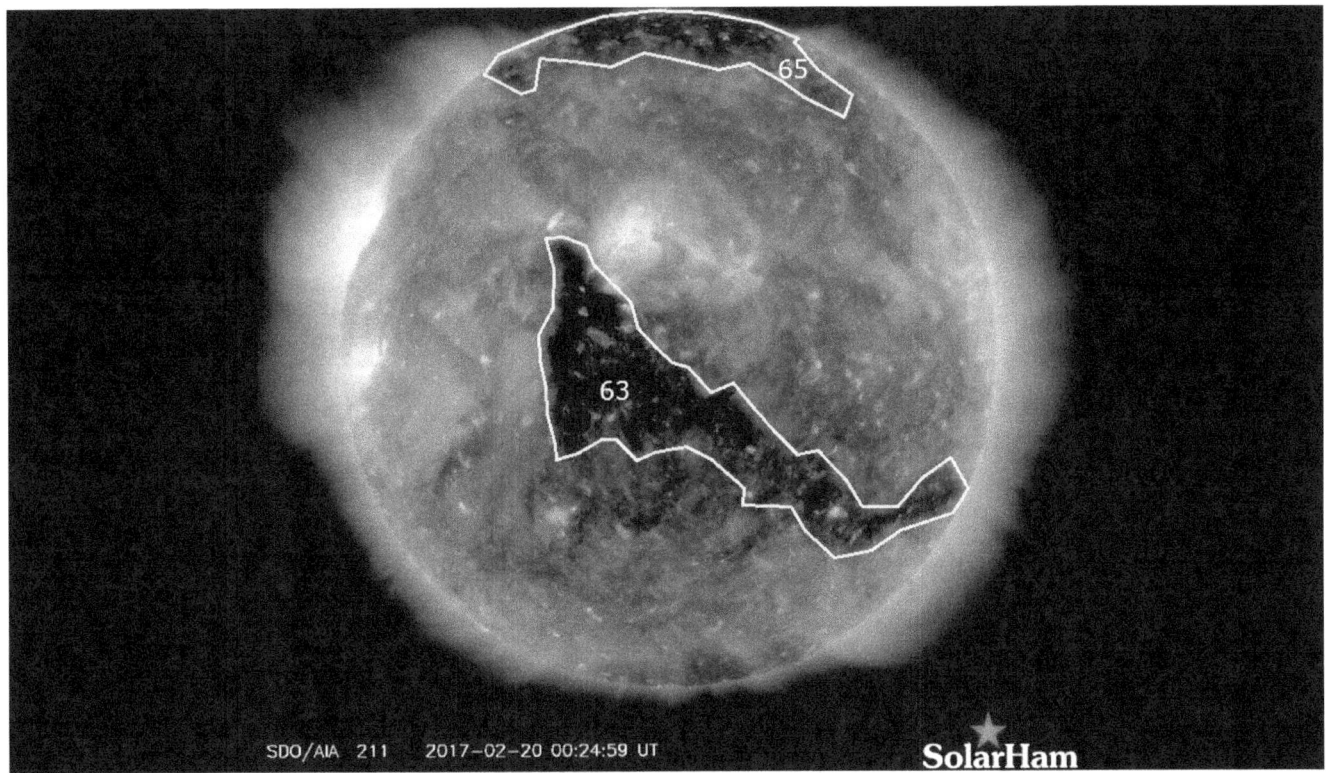

When an Earth-oriented coronal hole opens up on the Sun, which is a localized area of lower plasma density, larger volumes of solar cosmic-ray flux are able to escape the Sun and penetrate the atmosphere of the Earth. If this was the cause for the spikes in the experiment, then we should expect to see the repetitive spikes diminishing as the coronal hole on the Sun rotates away from the Earth with the rotation of the Sun.

Sun is by far the major contributor of cosmic-ray flux

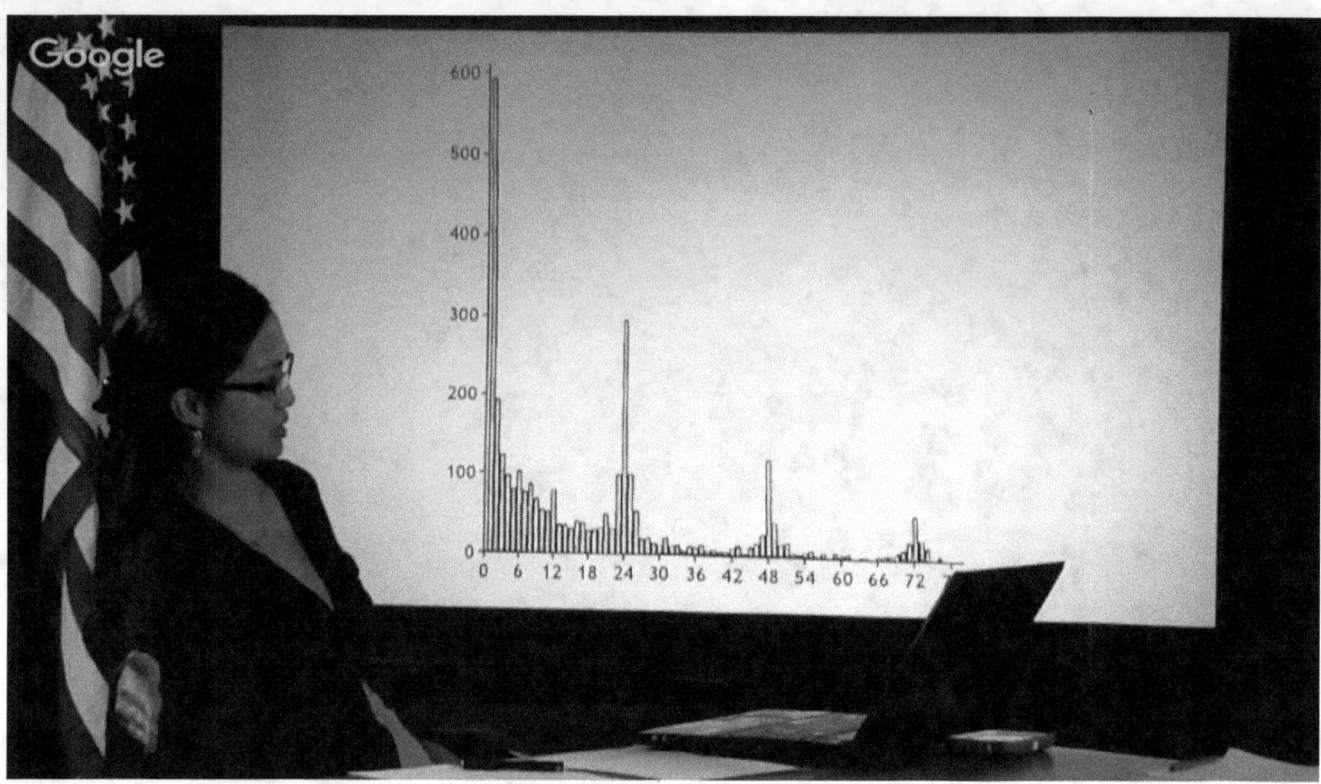

And this is precisely what we see reflected in the experiment.

If the cause for the spikes had been galactic, all the spikes would have been the same. If the cause was solar, then we should see the spikes diminishing, as we do.

The experiment proves thereby that the Sun is by far the major contributor of cosmic-ray flux in the Solar system, and thereby to the climate on Earth.

We have many examples that prove

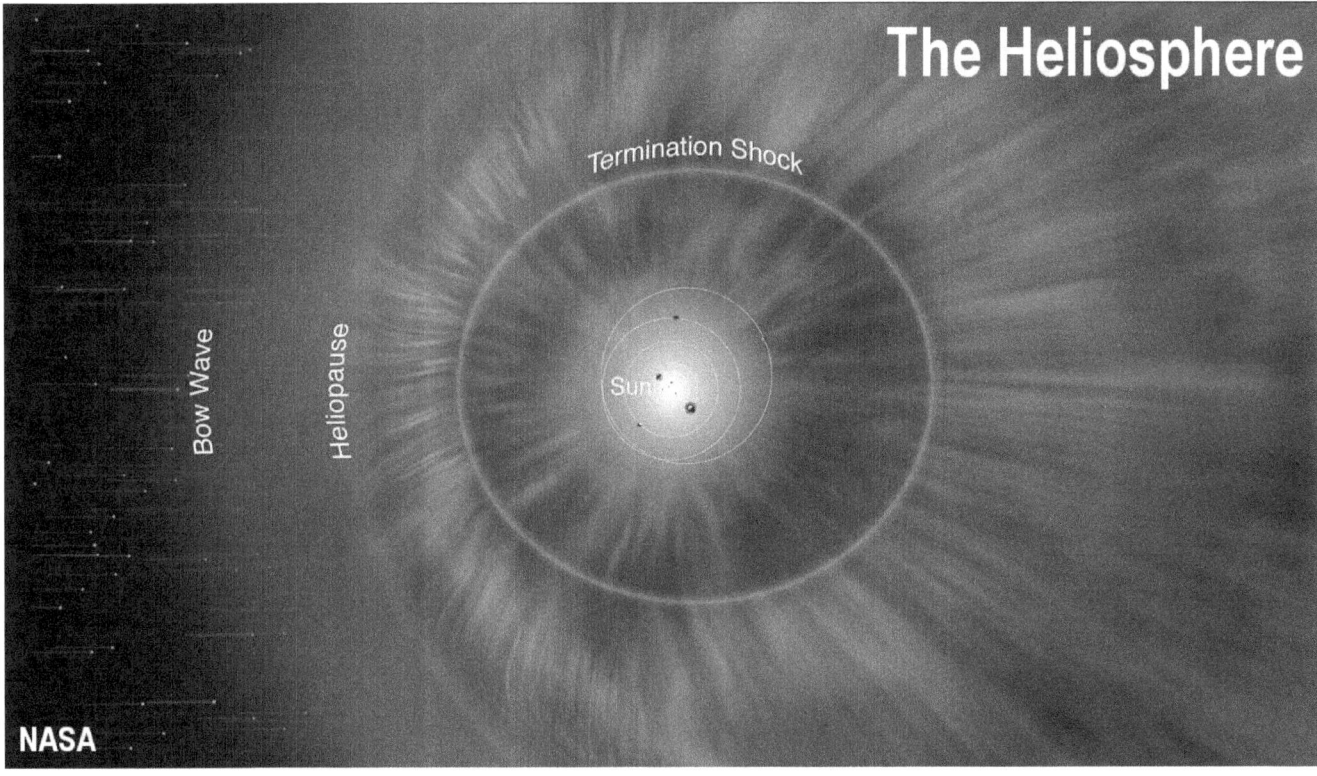

We have many examples that prove that the Sun is the major cause for cosmic-ray flux in the solar system.

How big a wallop the Sun packs

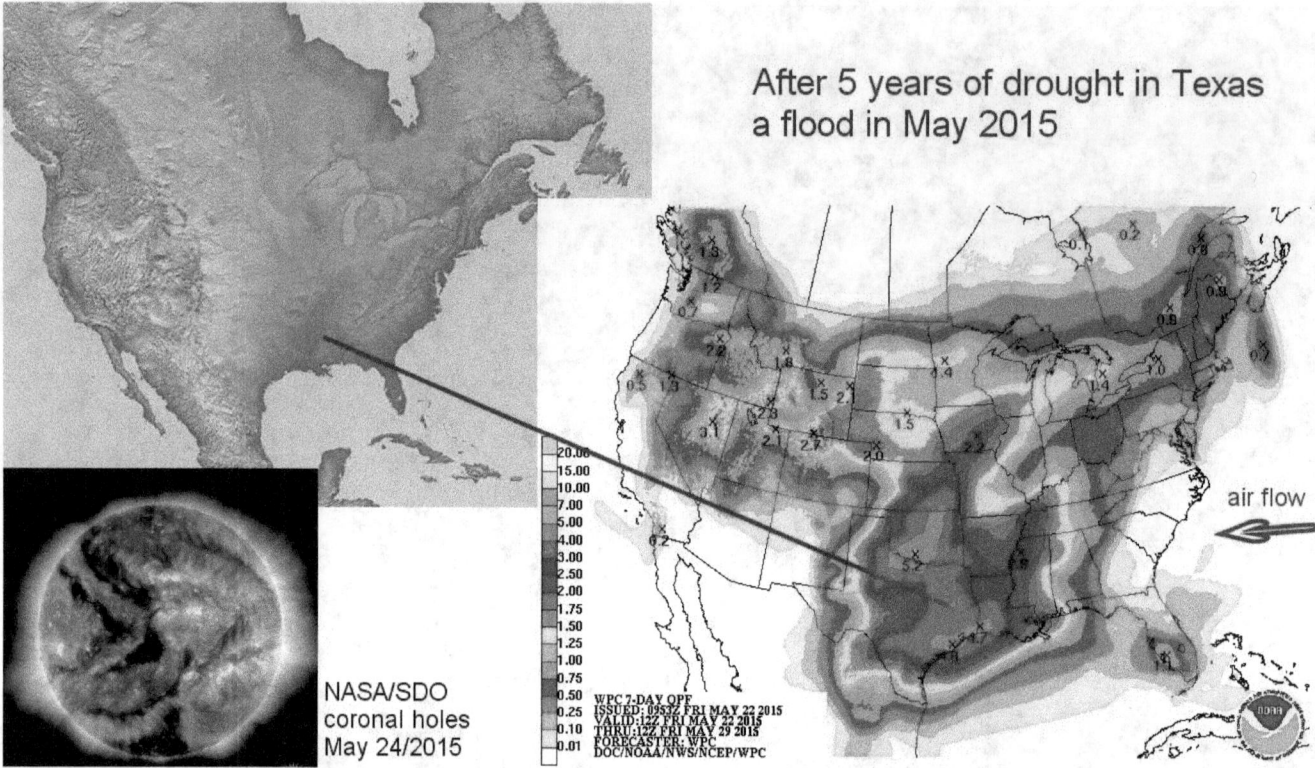

How big a wallop the Sun packs beneath its plasma corona becomes evident when a major hole opens up.

An example for such a potential occurred in 2015. After years of harsh drought conditions, a large flash-flood suddenly erupted in May 2015 that extended from Texas all the way to Canada. The flash flood event was likely caused by a major coronal hole opening up on the surface of the Sun that had briefly opened a window for an extreme volume of solar cosmic-rays reaching the Earth.

The Sun is evidently the major contributing factor

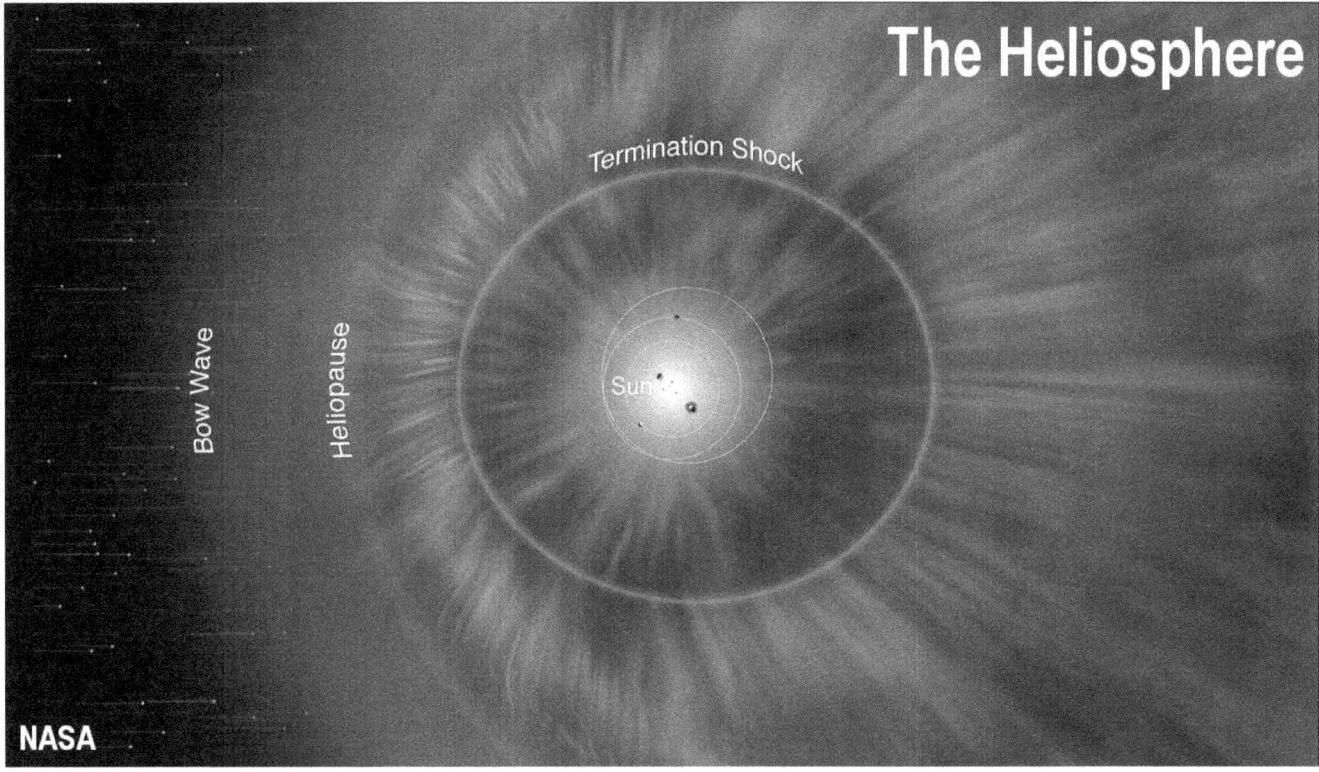

The Sun is evidently the major contributing factor for the historic Beryllium volumes that represent cosmic-ray flux affecting the Earth, which have been recorded on Earth including for the timeframe of the glaciation periods.

➤ Determining the start of the next Ice Age

**Determining the start
of the next Ice Age on Earth**

Determining the start of the next Ice Age on Earth.

The changing Sun is the driving factor for climate changes

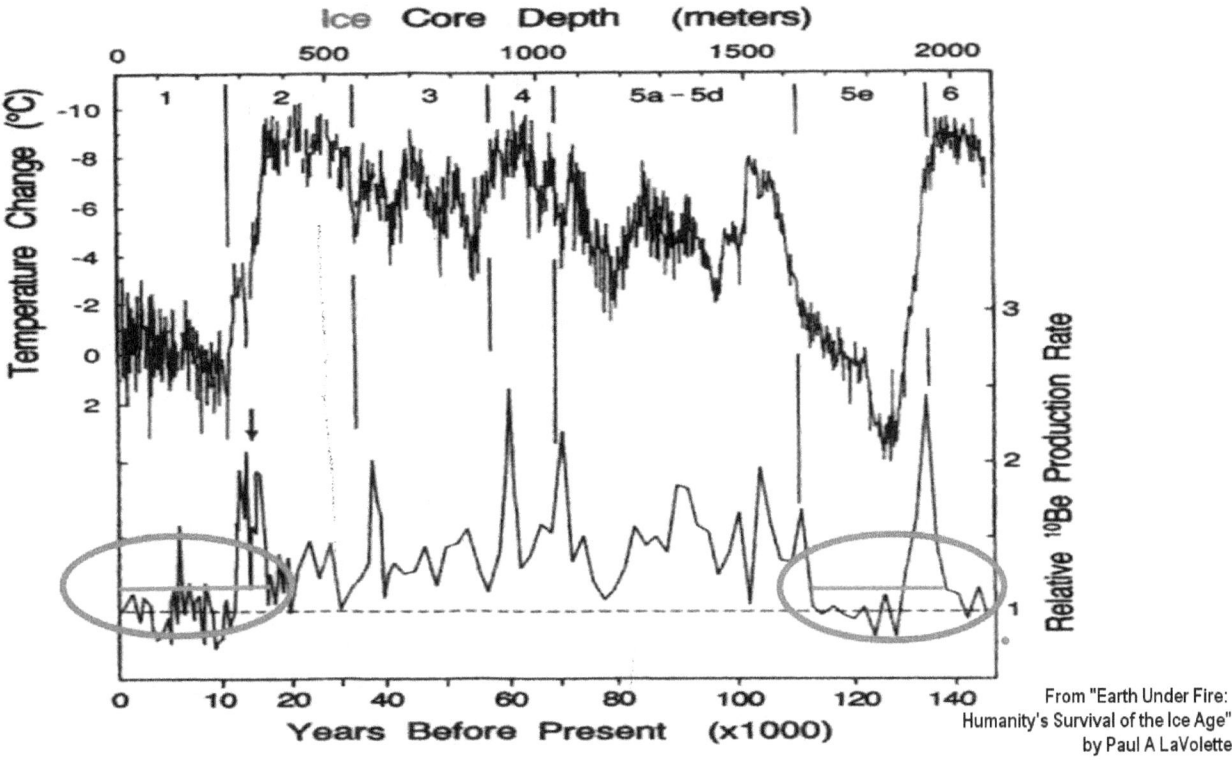

From "Earth Under Fire: Humanity's Survival of the Ice Age" by Paul A LaVolette

scientifically established recognition that the changing Sun is the driving factor for climate changes on Earth, also applies to the start of the Ice Ages. The historic Beryllium values provides us a measure to judge the potential end of the current interglacial holiday from the cold and the start of the next glaciation phase of the Earth under a hibernating Sun.

If one compares the Beryllium record for the timeframe between the start and end of the previous interglacial period, with that of the current interglacial that had started around 15,000 years ago, one will note that the current interglacial is already slightly longer in duration. This means that the startup of the next glaciation period is imminent.

Professor Dr. Zbigniew Jaworowski, has warned

"The Ice Age is Coming"
(2003 paper)

*Zbigniew Jaworowski, PhD, MD, D.Sc.
chairman of the Scientific Council
of the Central Laboratory
for Radiological Protection in Warsaw*

... a world-renowned atmospheric scientist and mountaineer, who has excavated ice out of 17 glaciers on 6 continents in his 50-year career.

The world renowned atmospheric scientist and chairman of the Scientific Council of the Central Laboratory for Radiological Protection in Warsaw, Professor Dr. Zbigniew Jaworowski, has warned in his 2003 paper, 'The Ice Age is Coming,' that the end of the current interglacial is already overdue by about 500 years, and he suggested that the transition could unfold rapidly, in potentially a single year.

This is what the Beryllium record also indicates

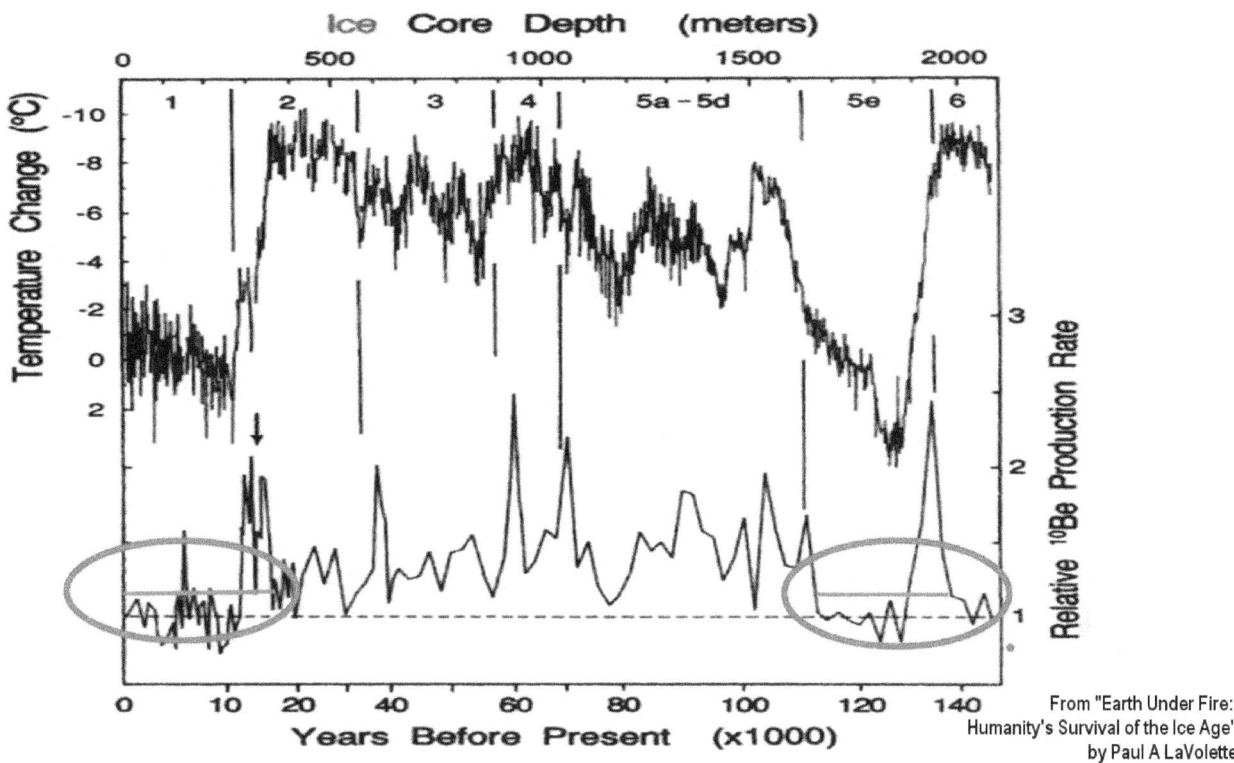

This is what the comparison of the Beryllium record also indicates. We even see supporting evidence measured in space.

The cosmic-ray increase Ulysses has measured

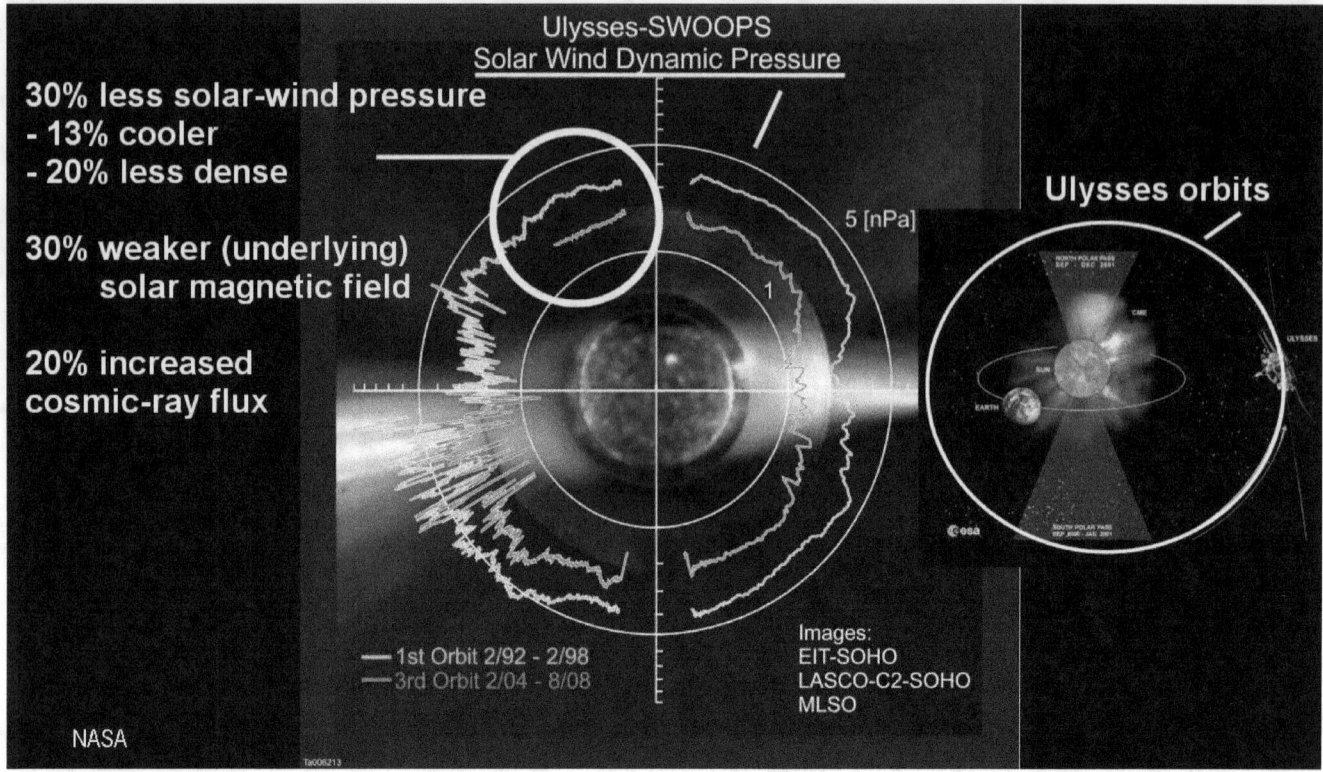

We see the supporting evidence reflected in the cosmic-ray increase that the Ulysses spacecraft has measured between 1998 and 2008, which saw the cosmic-ray flux increasing at a rate of 20% in 10 years. In other words, the transition to the next Ice Age is already actively in progress and is significantly advanced.

Ice Age already beginning, measured in space

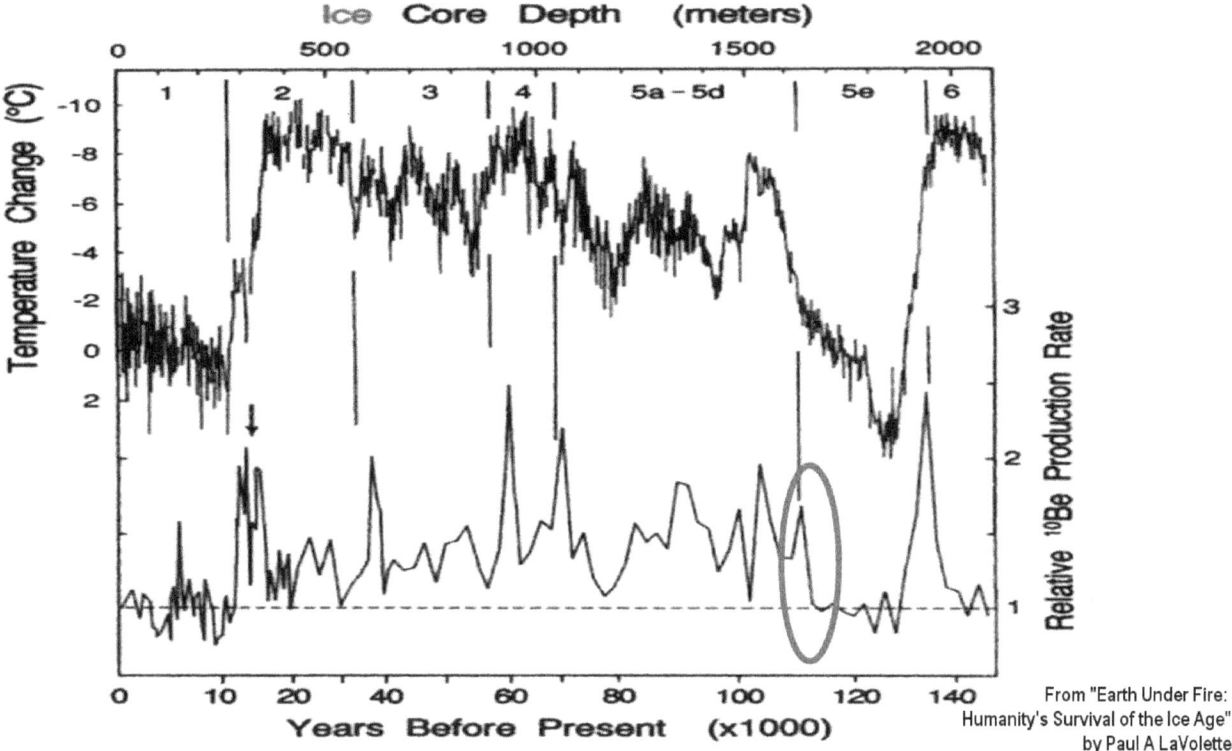

From "Earth Under Fire: Humanity's Survival of the Ice Age" by Paul A LaVolette

We see the equivalent of the Beryllium spike that started the last Ice Age already beginning, measured in space, caused by the weakening Sun, as the diminishing sunspot cycles also indicate.

That the galactic portion is minuscule

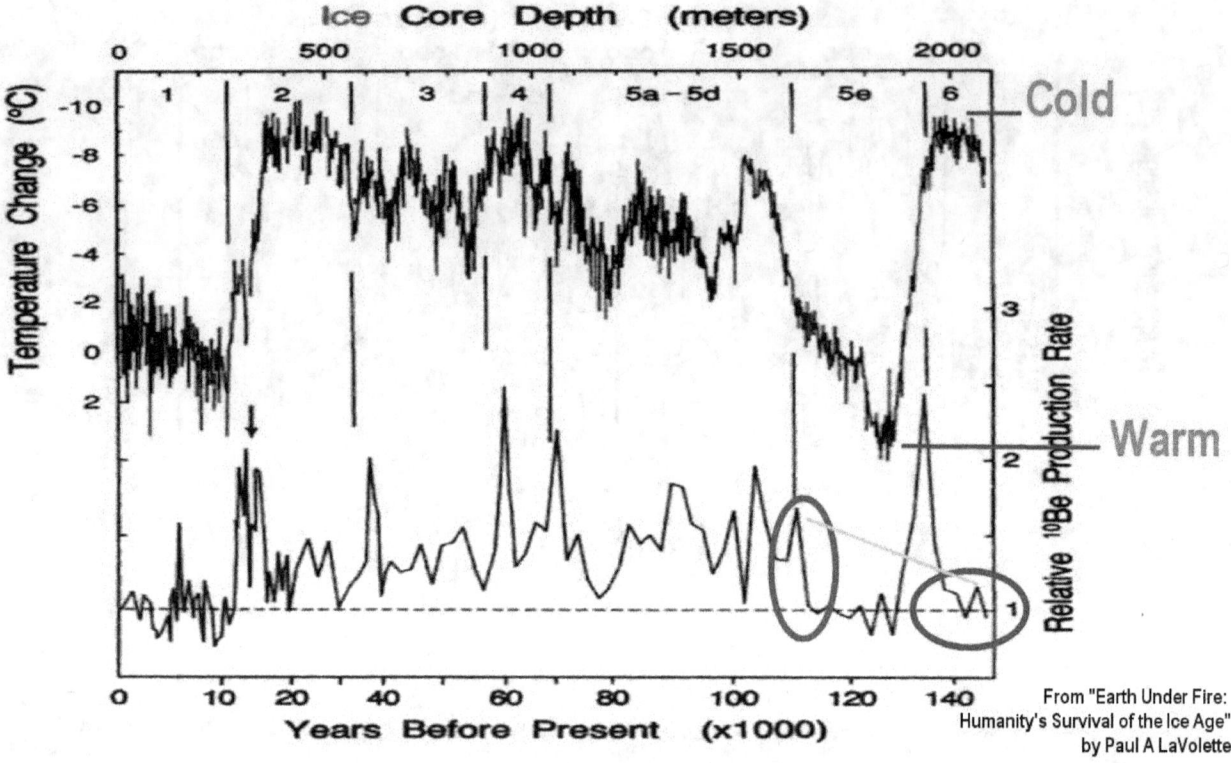

From "Earth Under Fire: Humanity's Survival of the Ice Age" by Paul A LaVolette

That the galactic portion is minuscule is evident by the extremely low Beryllium values past the 140,000-years mark. The Earth was in a deep glacial state at this time. The Sun was hibernating in a low-power state. The hibernating Sun does not have a dense plasma sphere around it that blocks solar cosmic-ray flux. This means that the low Beryllium value for this timeframe tells us, that this Sun was dramatically weak at the time, and also potentially dramatically cold.

My first reaction was that we should see a much higher level of Beryllium indicated at the end of the previous Ice Age. I thought that we should see something similar to what we see plotted for the glacial period from 110,000 years ago to about 20,000 years ago? I had expected to see similar high values plotted. Was the plotted low value for around 140,000 years ago, a mistake then?

I soon realized that the plot was not in error, but indicated that the hibernating Sun had diminished to extremely low levels of activity during the previous Ice Age, so that its solar cosmic-ray flux had diminished with it to extremely low levels, far below the start-up level for a glaciation period, as we had it 110,000 years ago.

This tells us that the solar activity, even during its low-power hibernation state, continues to diminish gradually throughout the glaciation period, ending at very-low levels and with extremely cold climates on Earth.

The low level that we see here at 140,000 years ago, reflects what I remembered from earlier research...

Living in caves at Pinnacle Point

...that the previous Ice Age that ended with such a harsh climate that according to archeological research, the human presence had dwindled from more than 10,000 adults to just a few hundred people remaining during the later portion of that glaciation period, all living in caves at Pinnacle Point on the shore of South Africa, nourished by the sea.

(By Curtis W. Marean - Scientific American magazine Vol.25, No4, Autumn 2016, p.37)

The critical point for us at the present time

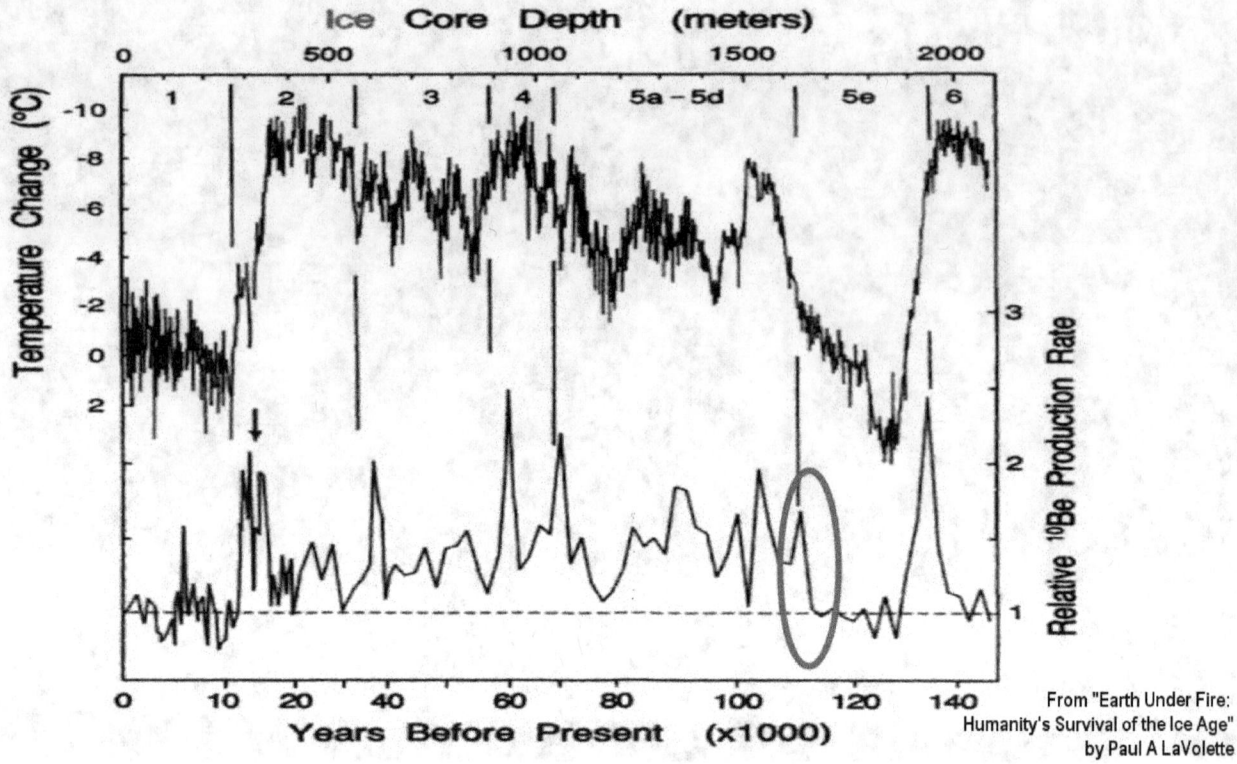

The critical point in this, for us at the present time, is not the extreme harshness of an Ice Age at the end of its period, but the conditions that we will find at the beginning of it.

The beginning level had been 40 times colder

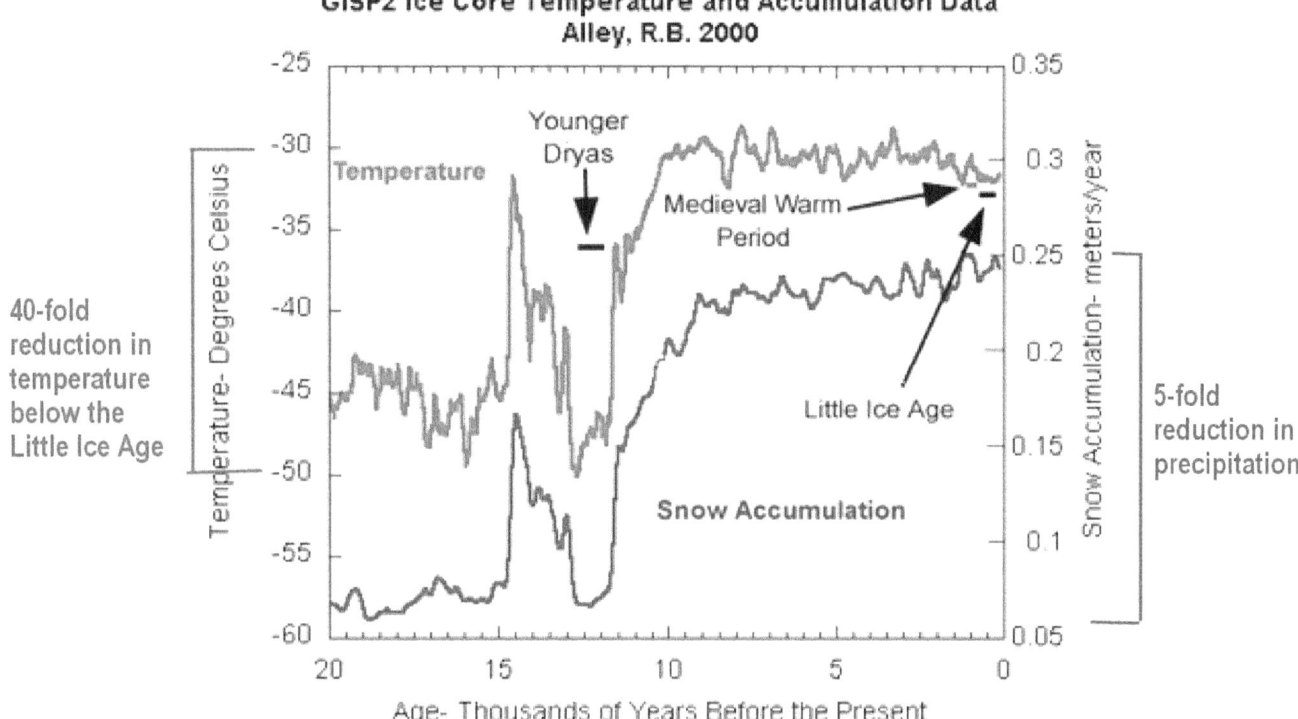

The beginning level promises to be roughly equal to the climate of the Younger Dryas period in which the climate had been 40 times colder than the cooling that was experienced during the Little Ice Age. This sets the stage of what we will experience again in the near term, potentially in the 2050s.

It is purely academic at this stage

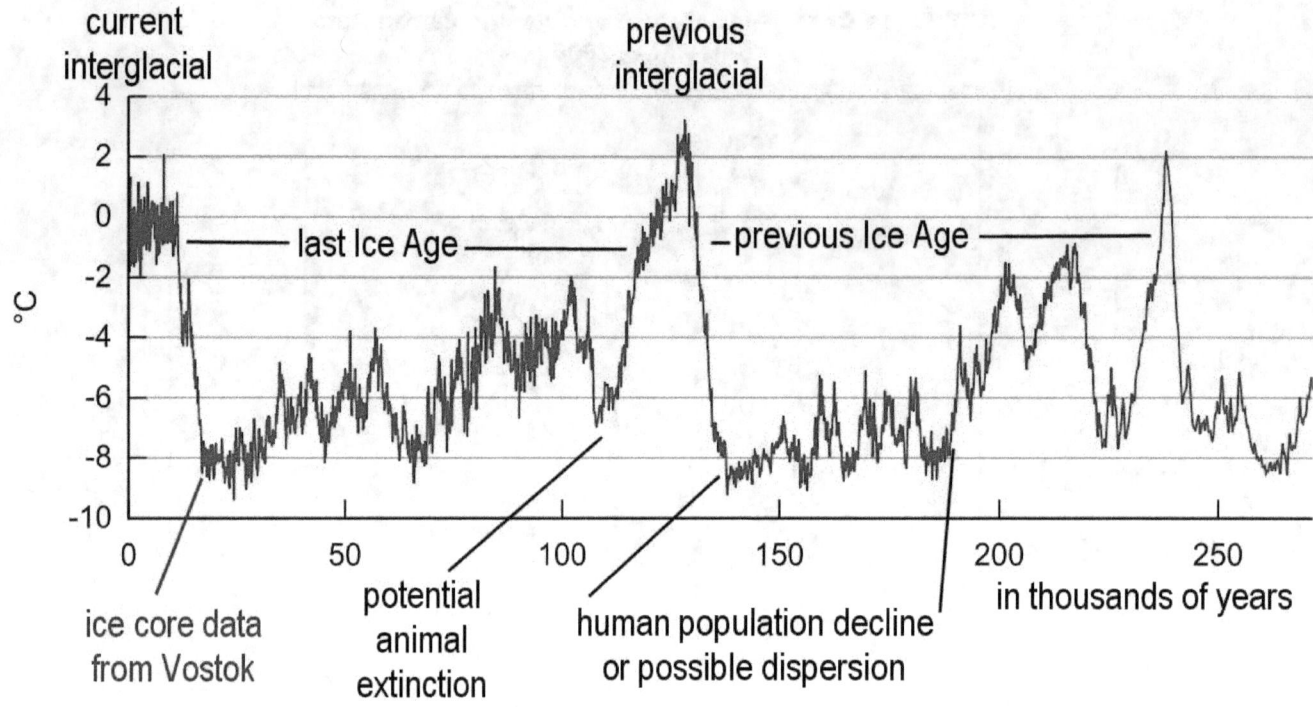

It is purely academic at this stage that the previous Ice Age appears to have been one of the harshest and had ended at an extremely low temperature, correspondingly with such a weak Sun that the beryllium levels were the lowest recorded for glacial conditions, and were significantly lower at the time, than the startup of the next Ice Age. The significance in this is our understanding of the dynamics of the processes and the principles that cause them.

When we get into glacial conditions

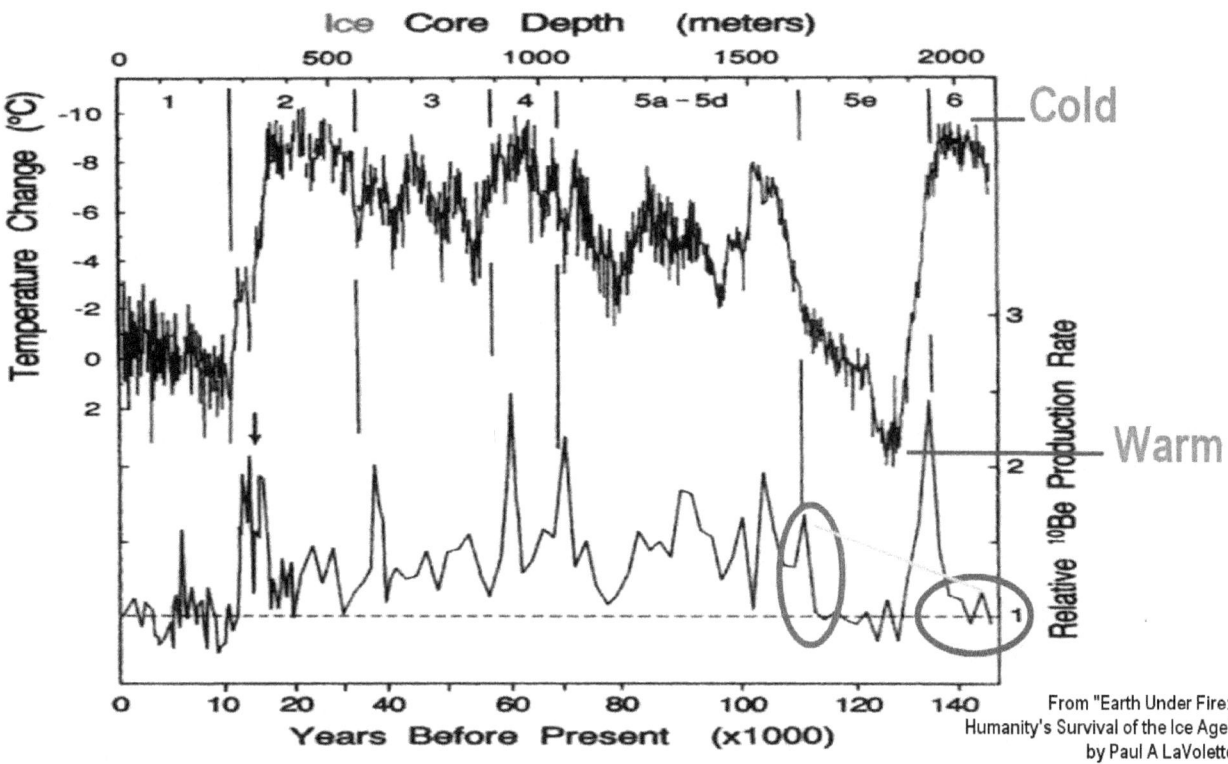

From "Earth Under Fire: Humanity's Survival of the Ice Age" by Paul A LaVolette

We have discovered that when we get into glacial conditions, the dynamics are rapidly reversed, because during glacial conditions, we are dealing with a weak, hibernating Sun that has almost no plasma sphere around it. When this hibernating Sun emits increasing cosmic-ray flux, this means that the hibernating Sun is getting more active and its surface temperature is getting hotter. Inversely, when we see diminishing cosmic-ray flux during the glacial period, the hibernating Sun is less active and its surface temperature is respectively colder. This is what the Beryllium measurements represent during glacial times.

When the Sun is in its high-powered active mode

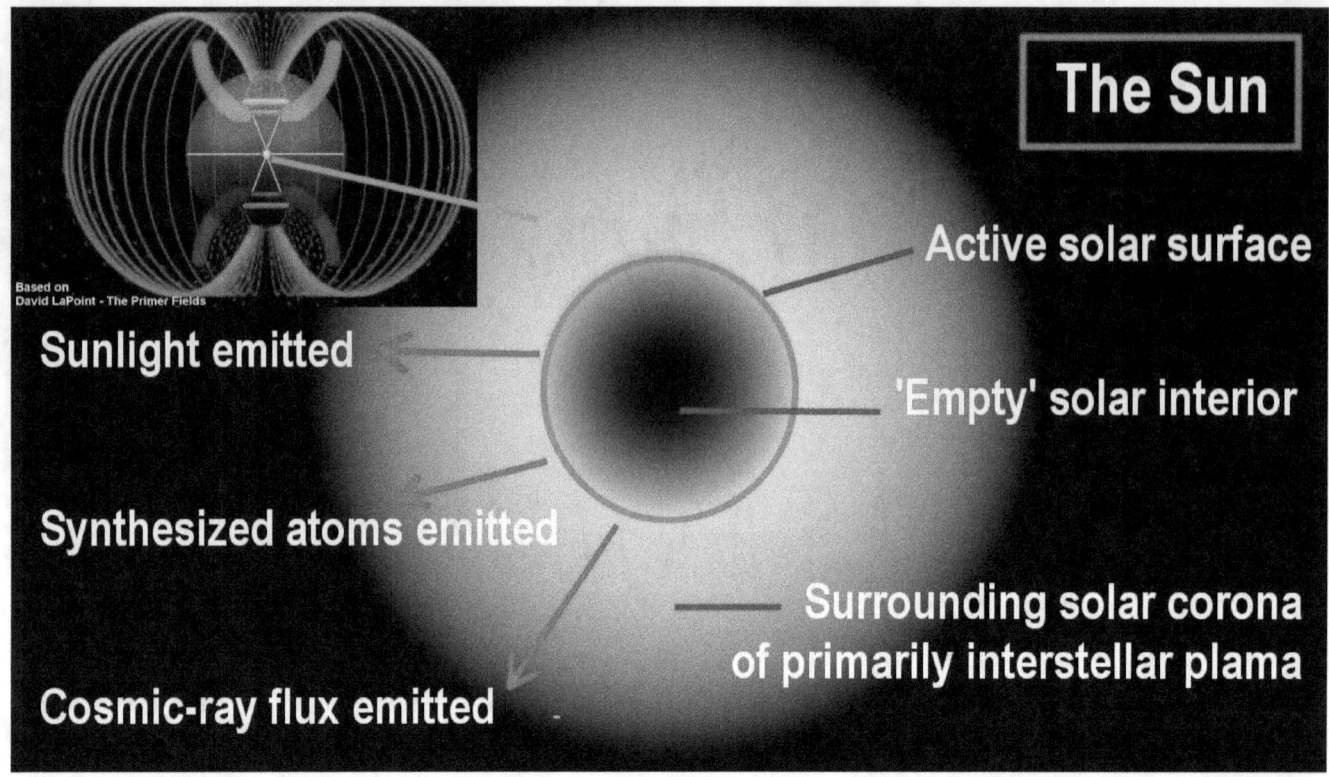

The complete opposite happens during interglacial time. When the Sun is in its high-powered active mode and is surrounded by a dense sphere of plasma that is focused on it by its electromagnetic primer fields, and such a Sun emits increasing cosmic-ray flux, then this means that this Sun is getting weaker. A weaker Sun has a weaker plasma sphere around it that enables a larger volume of cosmic-ray flux to escape through the plasma shielding.

When the sunspot numbers diminish for a weaker Sun

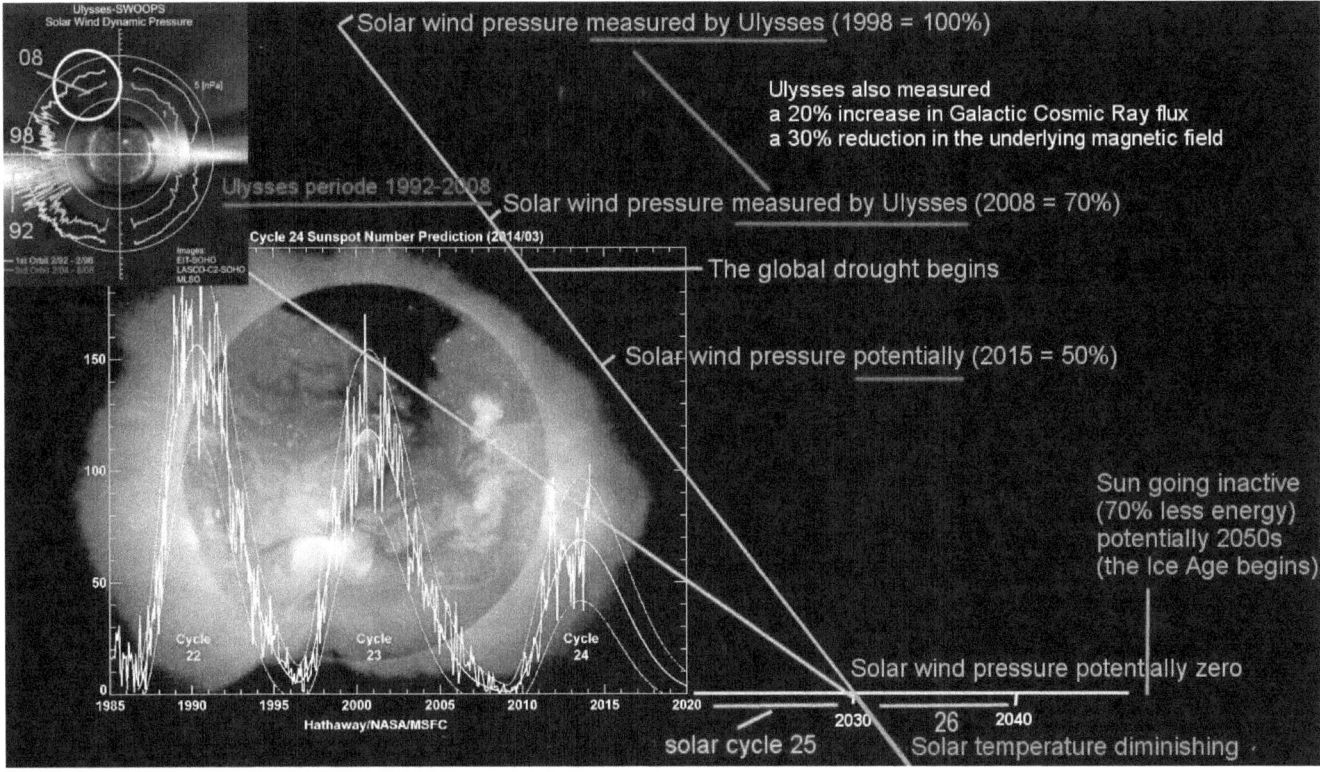

This is what happens in the background when the sunspot numbers diminish for a weaker Sun.

When this increase of the weakening happens rapidly, especially when the sunspots no longer appear, we are getting close to the next phase-shift to the Sun's hibernation state, and to the start of the next Ice Age.

This is precisely what we see presently happening. We see the weakening of the Sun also evident in the weakening of the solar wind pressure. All these factors are telling us that the solar cosmic-ray flux is now rapidly increasing, and the Sun is getting weaker.

We see signs of this rapid change all over the world in the form of increased global cooling, climate anomalies, increased droughts, floods, hurricanes, earthquakes, and so on.

➢ The Grand Solar Minimum - The Next Ice Age

```
┌─────────────────────────────┐
│                             │
│   The Grand Solar Minimum   │
│   ───────────────────────   │
│      The Next Ice Age       │
│                             │
└─────────────────────────────┘
```

The Grand Solar Minimum - The Next Ice Age.

The effect of the weakening of the Sun

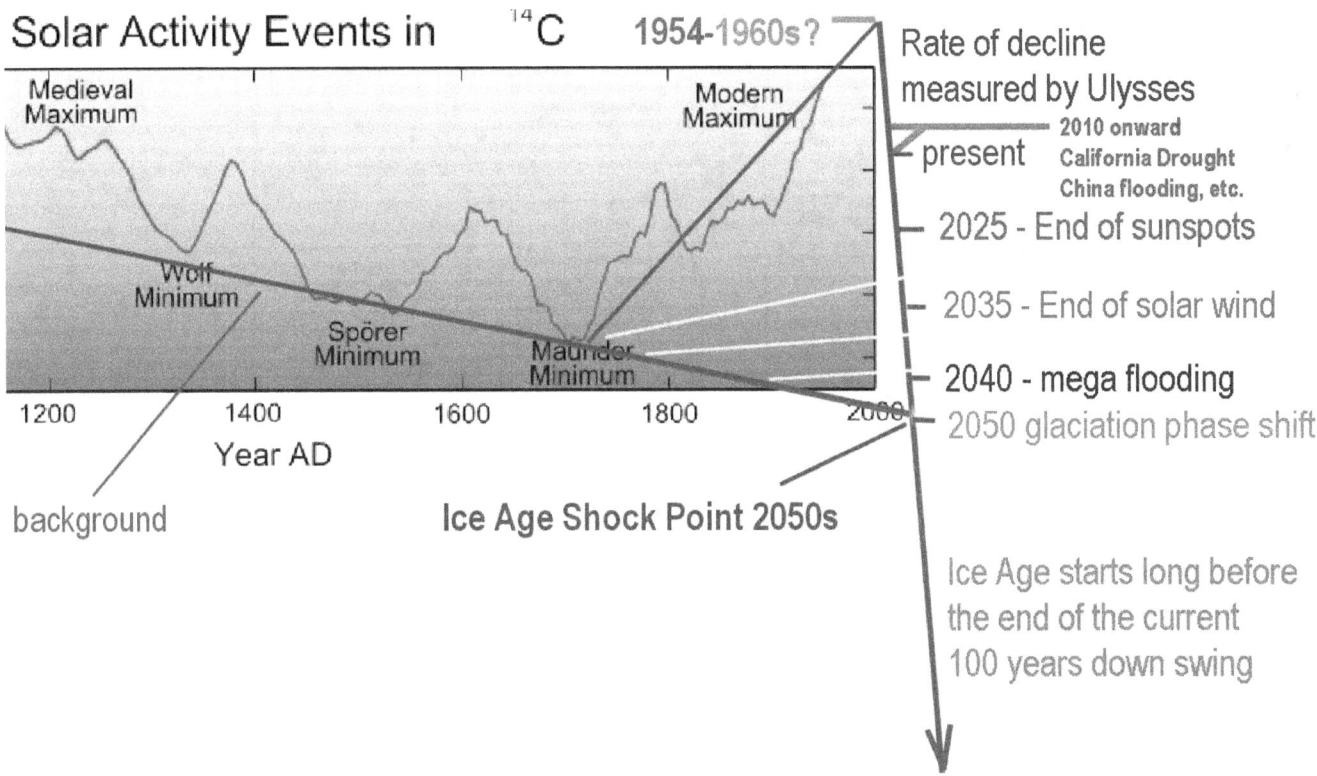

The effect of the weakening of the Sun is so dramatic now that one hears a lot of talk about another "Grand Solar Minimum" happening in the near future, such as the Maunder (grand) Minimum that gave us the Little Ice Age.

This concept of "The Grand Solar Minimum" is a deception, because the next Grand Solar Minimum will be the next Ice Age. The underlying support for the Sun no longer exists for a reversal to be possible from any Grand Minimum. The rapid collapse of the interstellar plasma density for the Solar System, some of which is evident in ever-greater coronal holes, has diminished the background for the Sun past the point for a possible reversal back to normal.

The Ice Age will be the next Minimum

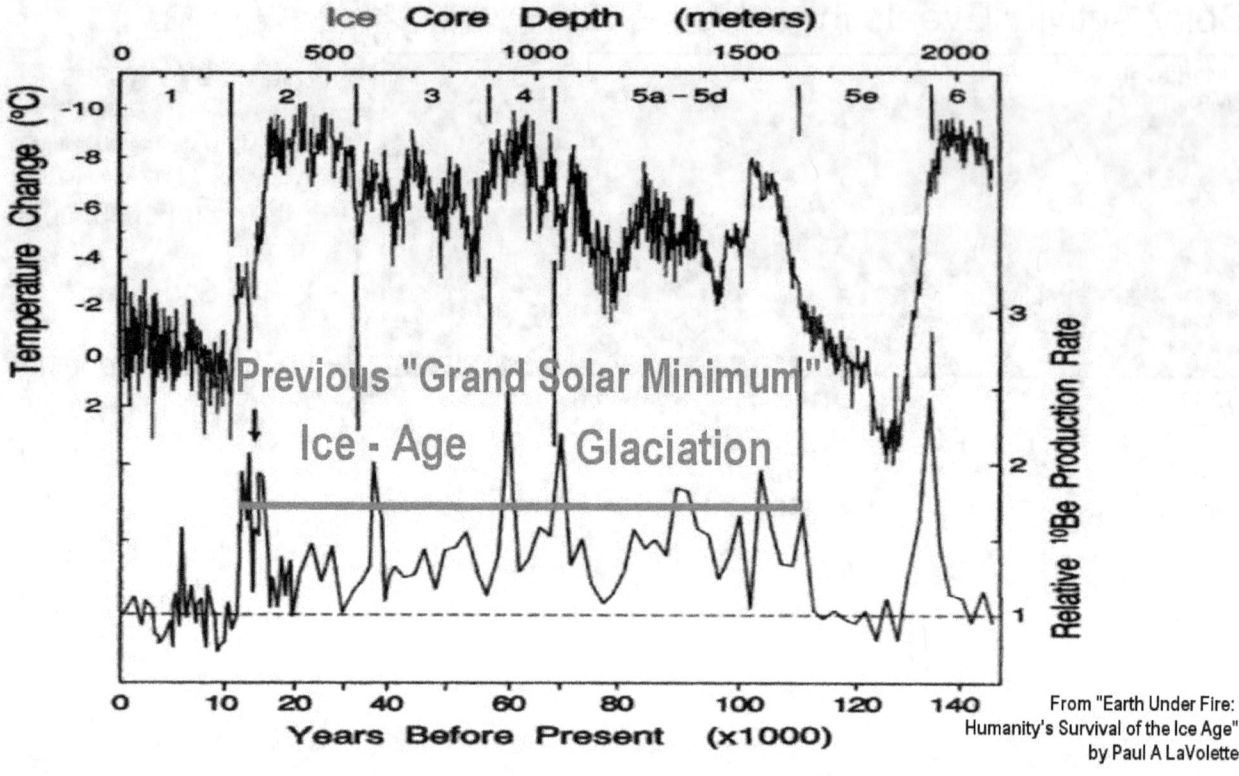

From "Earth Under Fire: Humanity's Survival of the Ice Age" by Paul A LaVolette

The Ice Age will be the next Minimum, and once the Ice Age has started with the collapse of the primer fields, it cannot be reversed without a major build-up in the interstellar plasma stream that is needed to rebuild the primer fields. This build-up happens of course, but will take a long time to develop. This means that the recovery from the now near "Grand Solar Minimum" that will be the Ice Age itself, will take about 90,000 years.

During the cold period of the hibernating Sun

View of the Earth from ISS, Jan.4 2013, from over the mid-Pacific, 460 miles east of northern Honshu, Japan.

During the cold period of the hibernating Sun, the relationship between high solar cosmic-ray values and colder temperatures continues to some degree, and continues to modulate the cold Ice Age climate by the effect of the still ongoing cosmic-ray flux on cloud forming. Increased cloudiness continues does reflect larger portions of the in-coming solar energy back into space.

During the glacial period, high Beryllium values

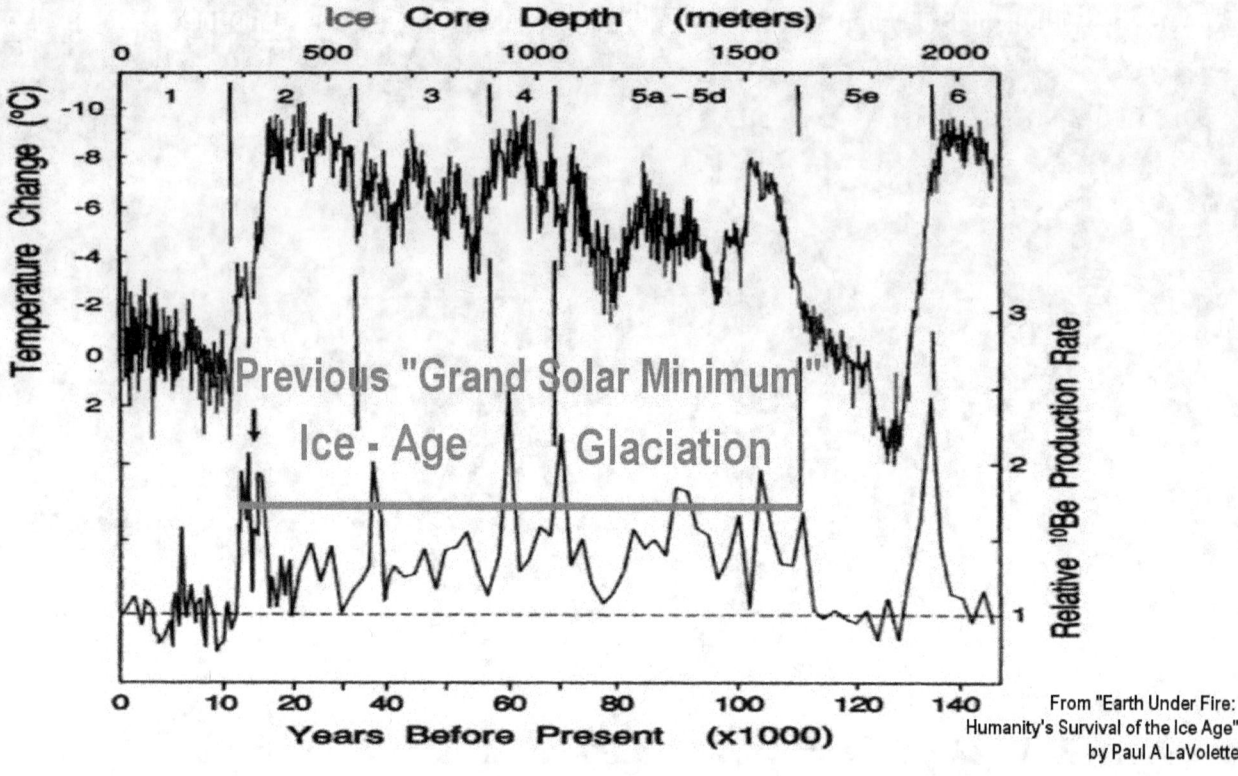

From "Earth Under Fire: Humanity's Survival of the Ice Age" by Paul A LaVolette

During the glacial period, high Beryllium values, continue to correspond with cold temperatures. This relationship indicates that cloudiness continues on the Earth, and that it continues to be affected by cosmic-ray ionization, and continues to be a major climate factor, although this may be in the form of ice fog resulting from the extreme cold.

It is unknown what the hemispheric distribution of the cloudiness will be during glaciation conditions. The ice fog may extend from the poles to close to the tropics, with the tropics being the rain belt, sparse as the rain may be.

We are not at the phase-shift point yet

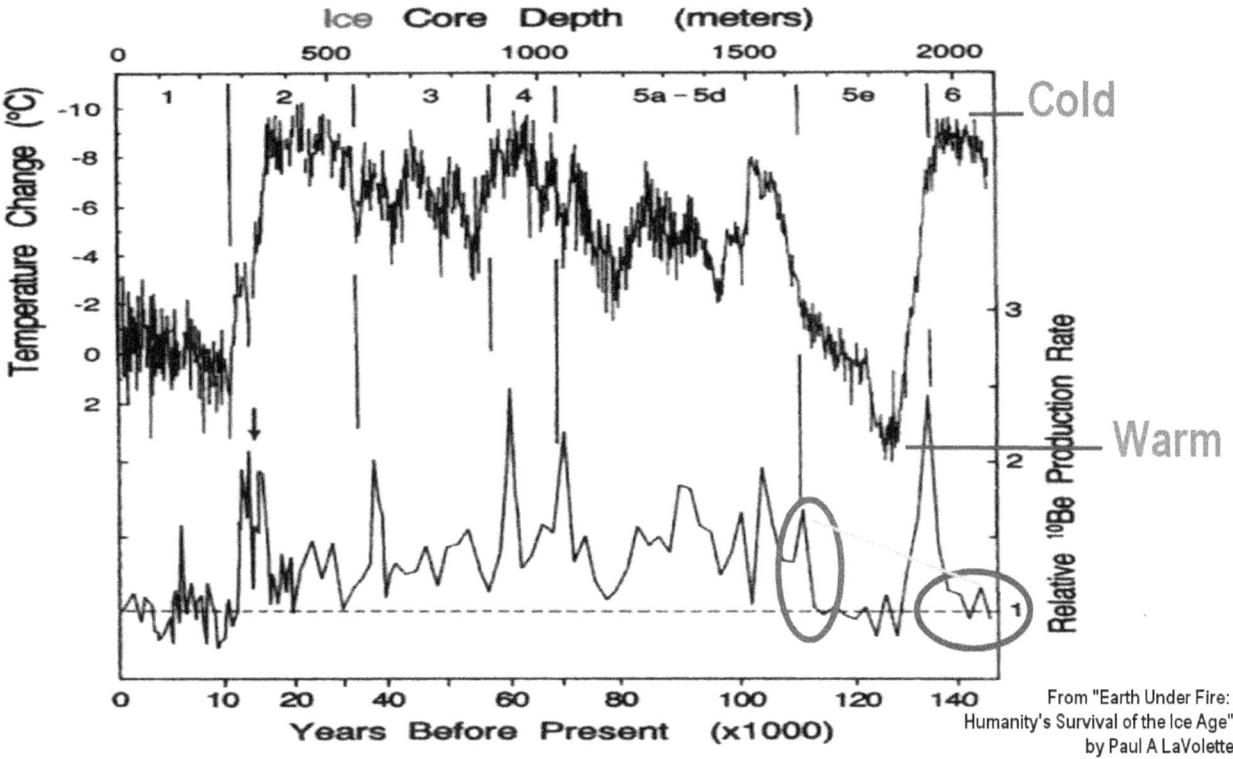

We are not at the phase-shift point yet where the glacial conditions begin as they did for the start of the last Ice Age, but the dynamics are progressing towards it with a phase shift becoming likely occurring in the 2050s.

The Little Ice Age in the 1600s

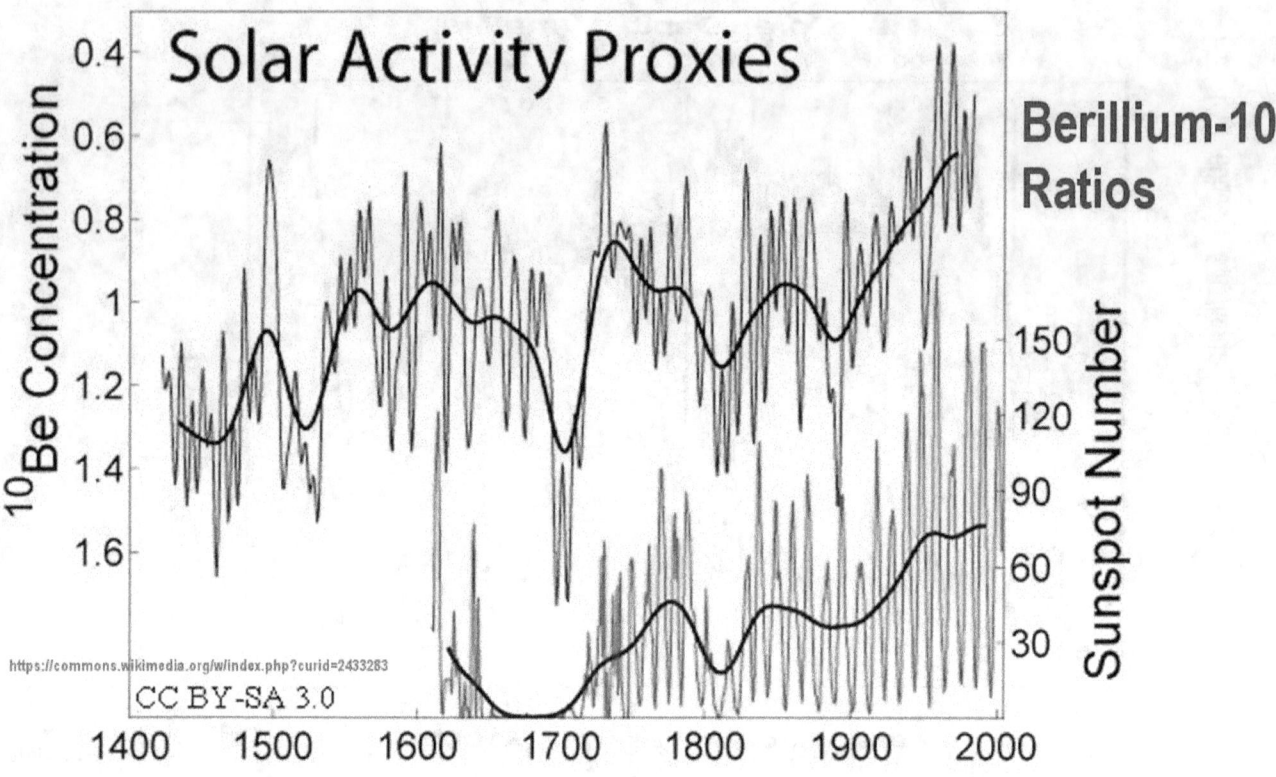

The Little Ice Age in the 1600s gave us a major peak value in cosmic-ray flux, according to the Beryllium-10 measurements. This trend towards colder climates may have brought us close to the phase shift. Fortunately for us, this trend towards the phase shift was interrupted in the 1700s by an uplifting pulse in the plasma system with interlocked resonance features that occur in long intervals and will not occur again soon.

If the up-lifting of the Sun had not occurred

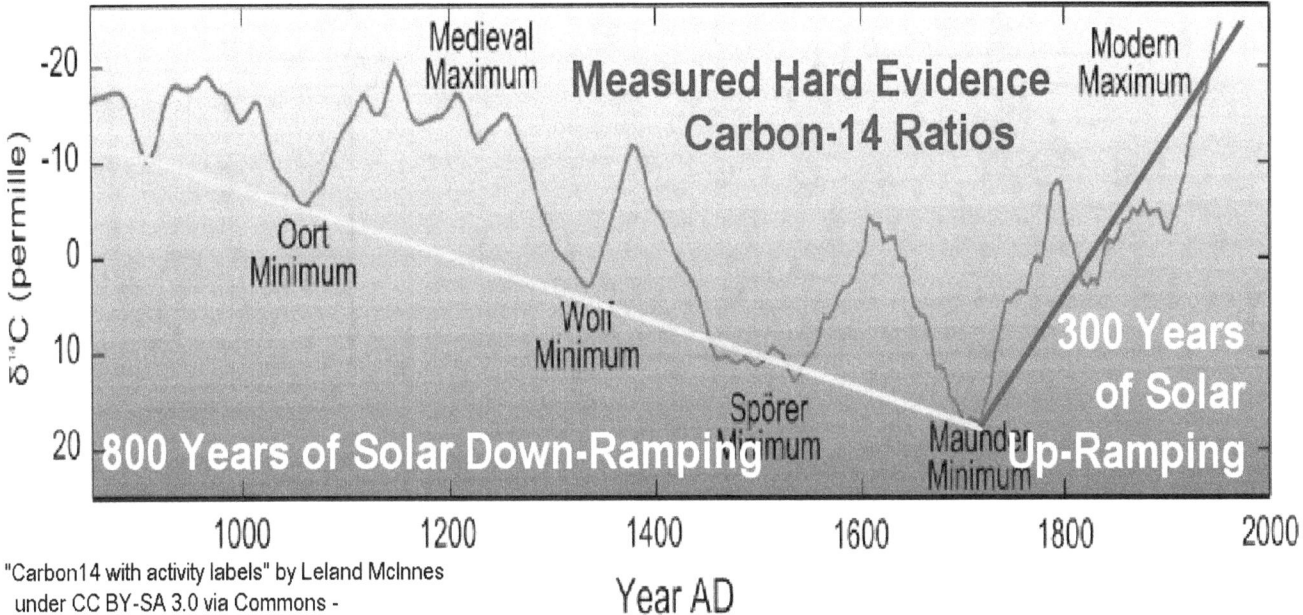

If the up-lifting of the Sun had not occurred in 1715, the majority of the people living today would not exist. We were not prepared for an Ice Age startup. Neither are we prepared now.

The rescue pulse that had prevented the Ice Age startup was followed by a strong recovery of the Sun. The increased solar activity gave us almost 300 years of global warming with dramatically reduced cosmic-ray-flux along the way.

The warming trend (1715-1998) is reflected

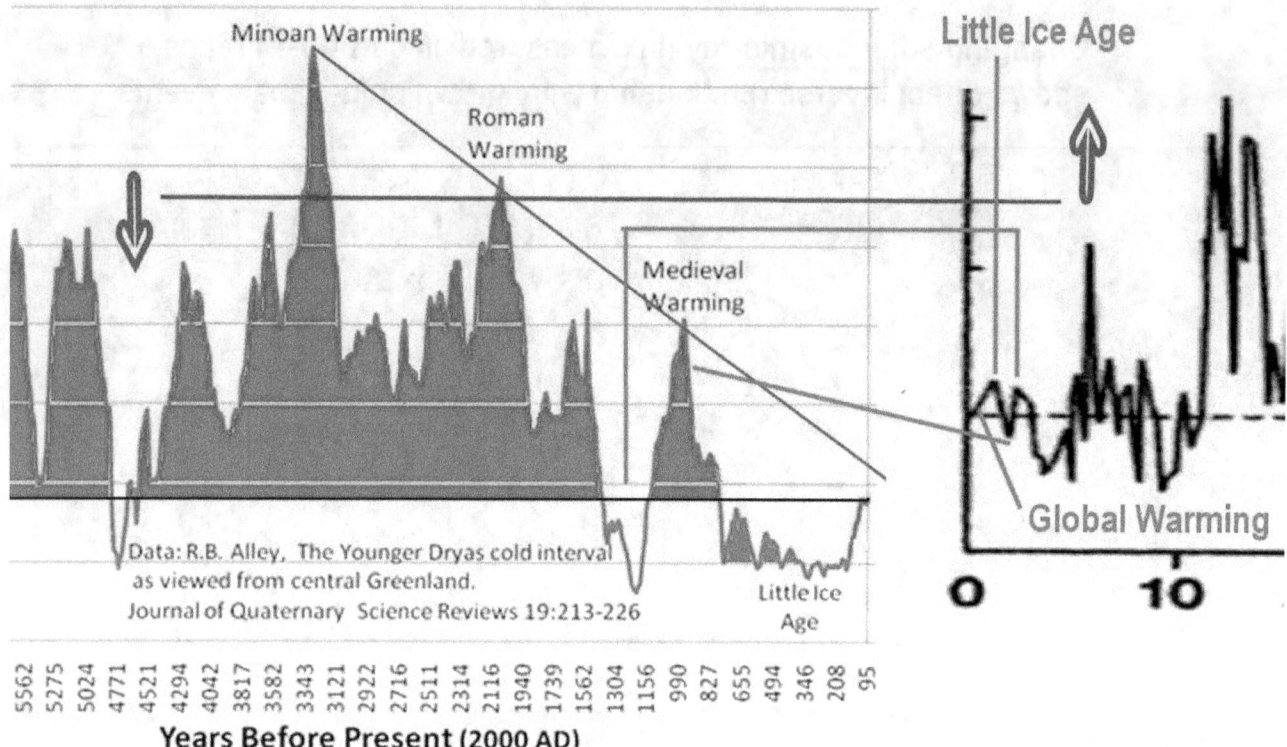

The warming trend (1715-1998) is reflected in diminishing Beryllium measurements that reflect the strengthened Sun.

The Sun became up-ramped by a plasma resonance pulse

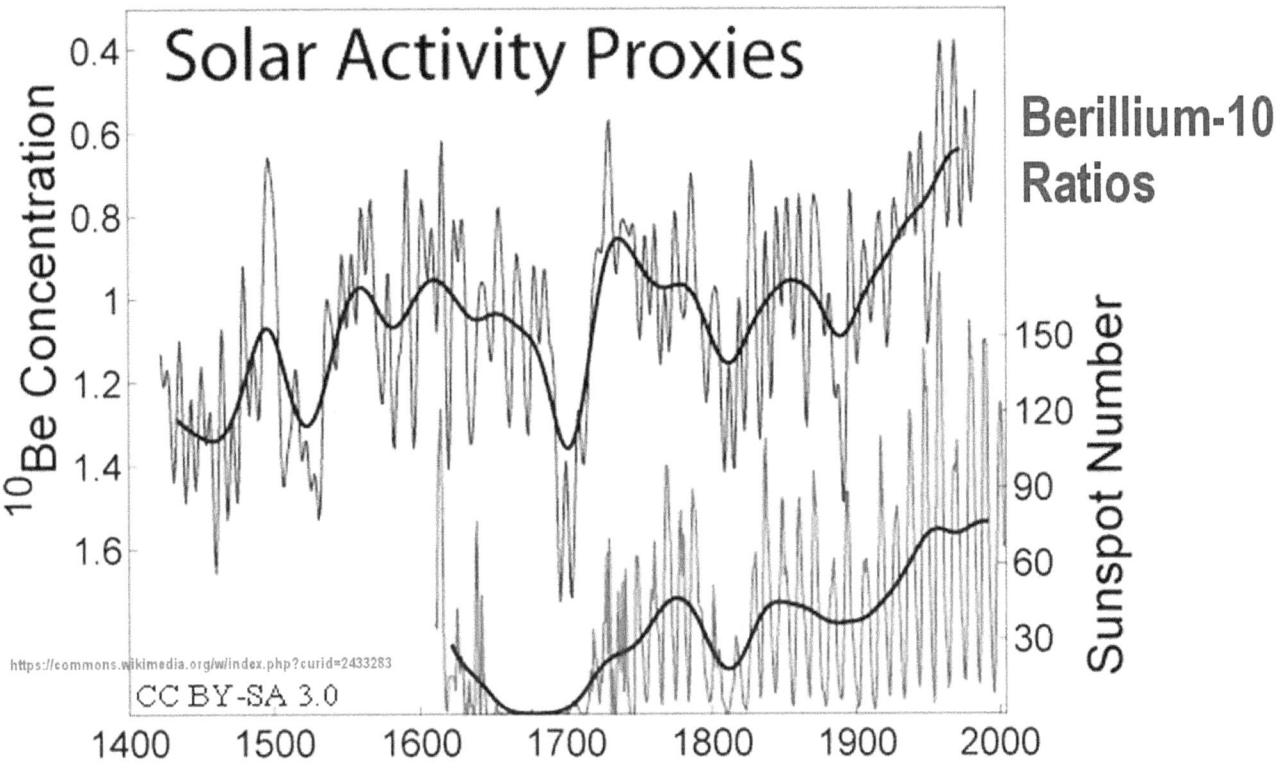

We see this critical period presented here in an expanded view. In this particular graphic the Beryllium record is inversed for easier correlation with temperature trends. By this reversal the warm climate of the stronger Sun, coincides with reduced cosmic-ray flux and low Beryllium measurements.

What we see here represents the global warming trend. It was caused exclusively by the Sun from the 1700s on. The Sun's becoming stronger is also reflected by increasingly larger solar activity cycles, the sunspot cycles.

This illustration shows that the infamous global warming that has been turned into a political scare campaign, had actually been caused by Sun. The Sun became up-ramped by a plasma resonance pulse that affected the solar system.

The warming pulse occurs in long intervals

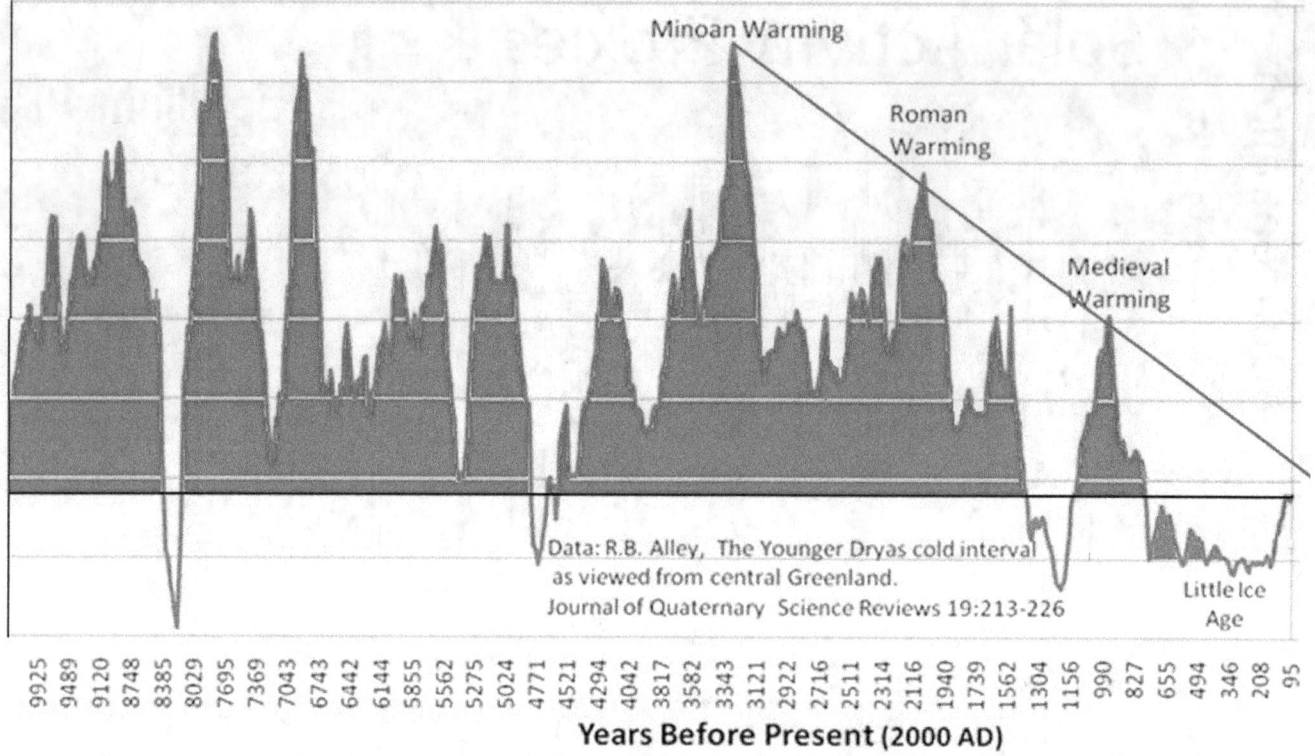

The warming pulse occurs in long intervals. The long interval places the next one outside the timeframe for the potential Ice Age phase shift in the 2050s. This means that the now unfolding phase shift will happen this time, without fail. Are we ready for it?

Also plotted in ratios of Carbon-14

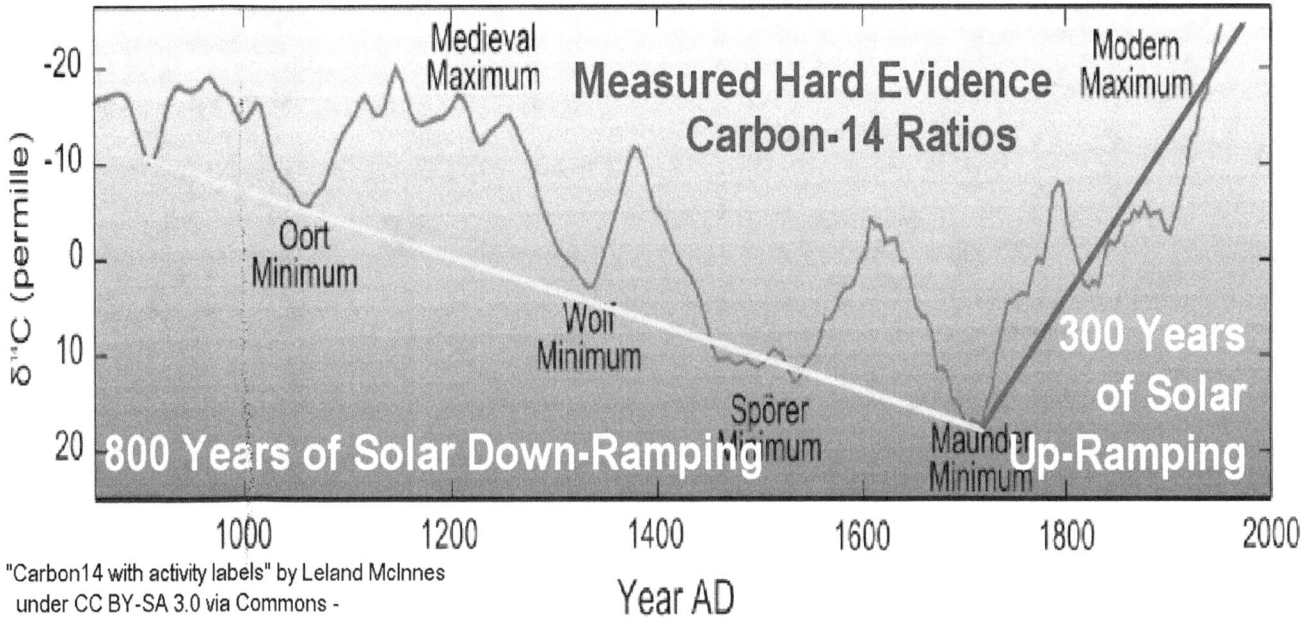

We can see the same solar dynamics also plotted in ratios of Carbon-14 measurements. Carbon-14 is another radio-isotope that is produced by cosmic-ray flux colliding in the atmosphere. At least it was so produced exclusively until the 1950s. In the 50s the Atomic bomb testing added another source for Carbon-14, which rendered the measurements thereafter useless as a solar activity proxy.

The space race that gave us the Ulysses satellite

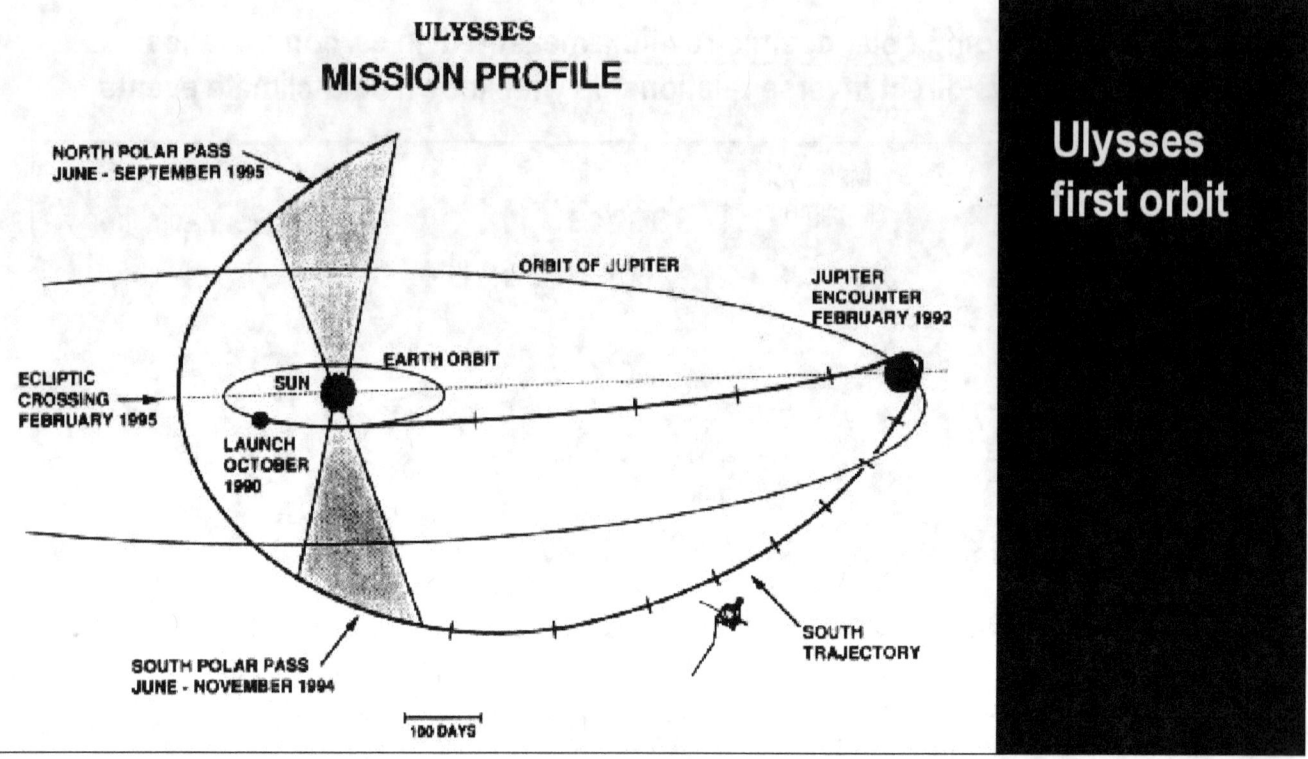

Fortunately, the arms race that closed down both radioisotopes window to the Sun, opened a new window with the space race that gave us the Ulysses satellite. In 1990 NASA launched this satellite that would observe the Sun from a polar orbit and measure its parameters directly as they happen.

Ulysses observed the Sun for 16 years

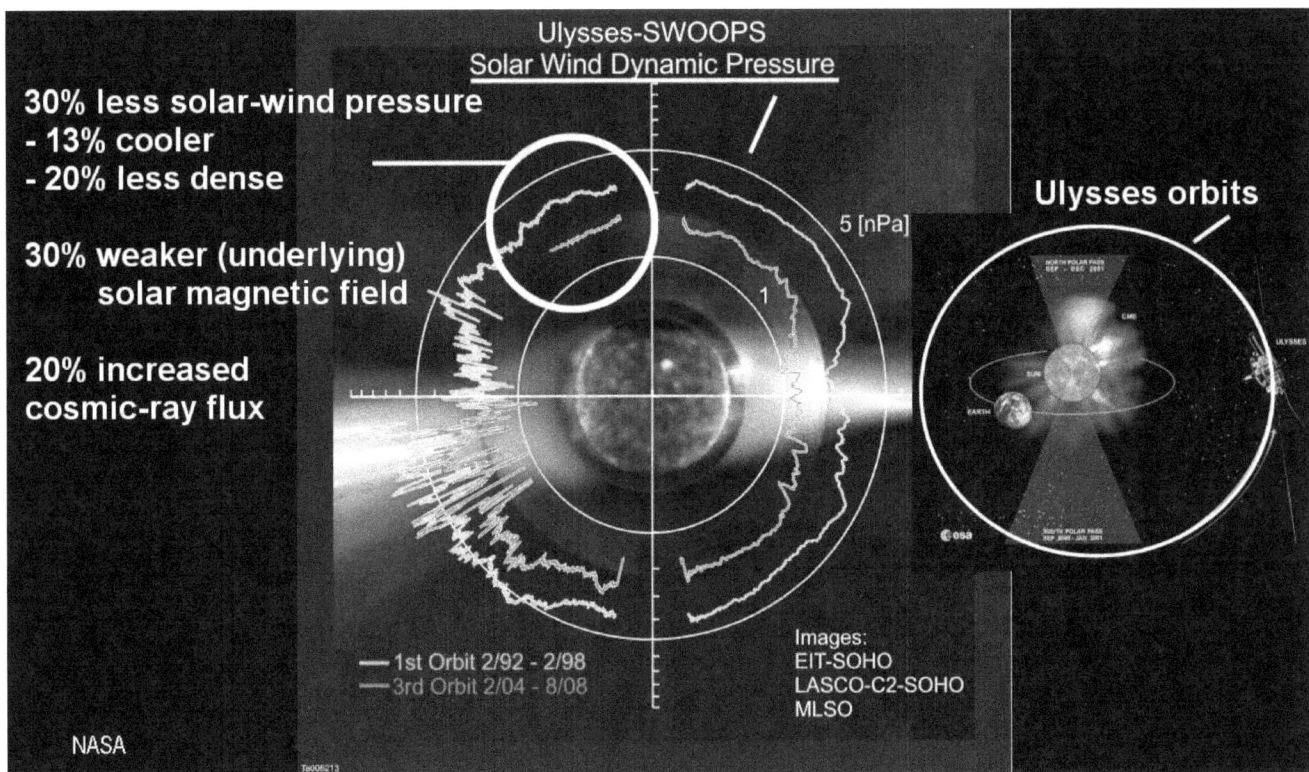

Ulysses observed the Sun for 16 years. It measured the Sun getting weaker. It saw the solar wind and the Sun's magnetic field diminished by 30% over 10 years, and the cosmic-ray flux increased by 20%. These are big numbers for such a short time on the cosmic scale. The importance of these numbers is that the numbers confirm that our Sun is rapidly diminishing, which had already been noted in the 1970s when the sunspot numbers had diminished dramatically.

Suddenly the up-trend was reversed in the 1970s

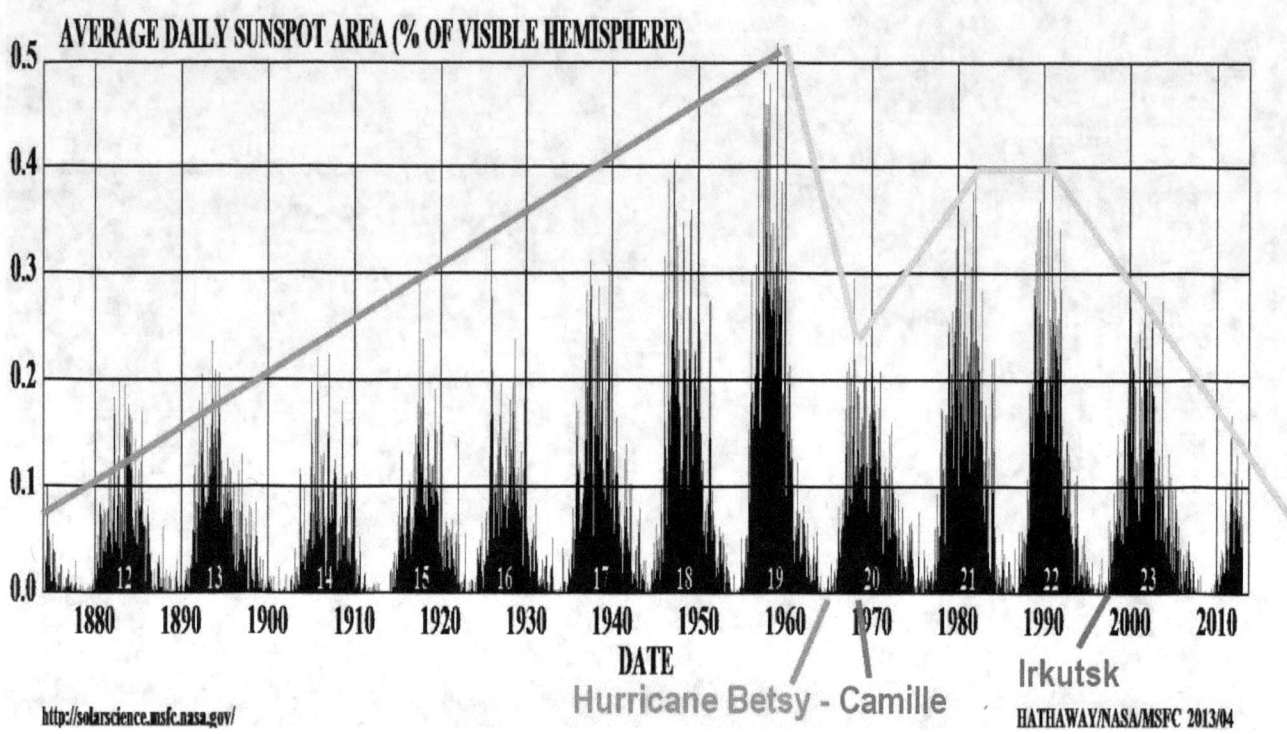

The sunspot numbers had been steadily increasing till the 1960s. Then, suddenly the up-trend was reversed in the 1970s. The reversal tells us that in the 1970s the big solar global warming may have peaked and the Sun was getting weaker again. We saw some big hurricanes in those days. The 1970s, in which the sharp reversal of the solar trend occurred, was also the time in which the Ulysses mission was perceived and engineered, which would eventually take 36 years to run to completion.

On-the-ground soil-temperature measurements

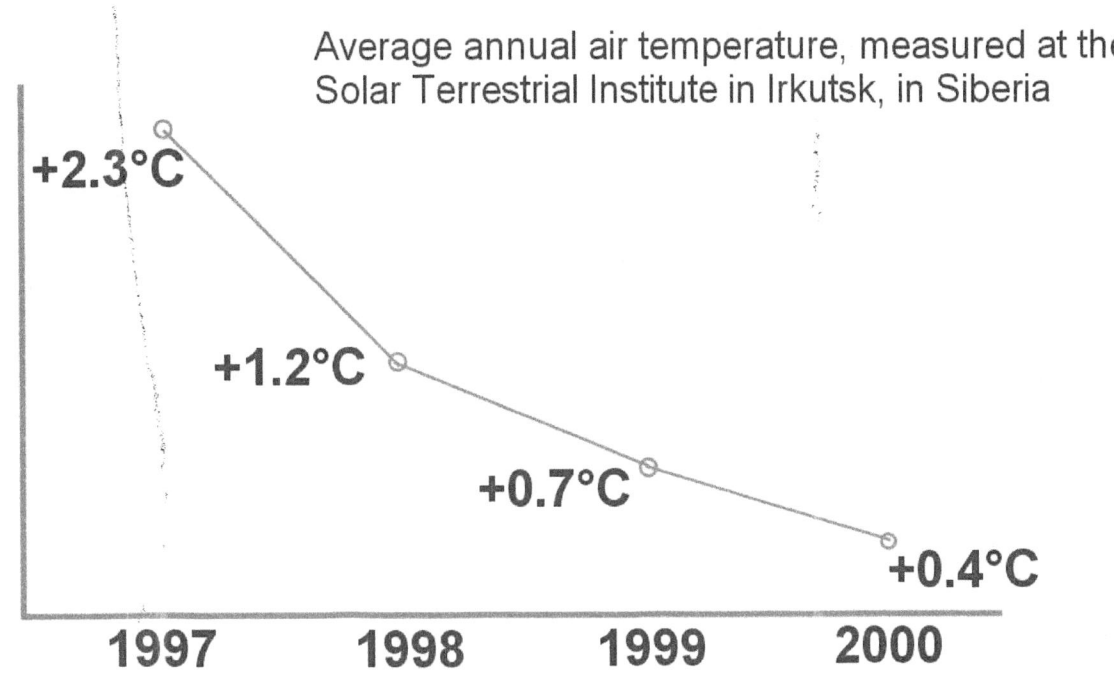

The first real temperature measurements that confirmed that the global warming trend had become history, were a series of on-the-ground soil-temperature measurements conducted in Irkutsk in Siberia, by Russia's Solar Terrestrial Institute.

Ulysses witnessed the great historic phase shift

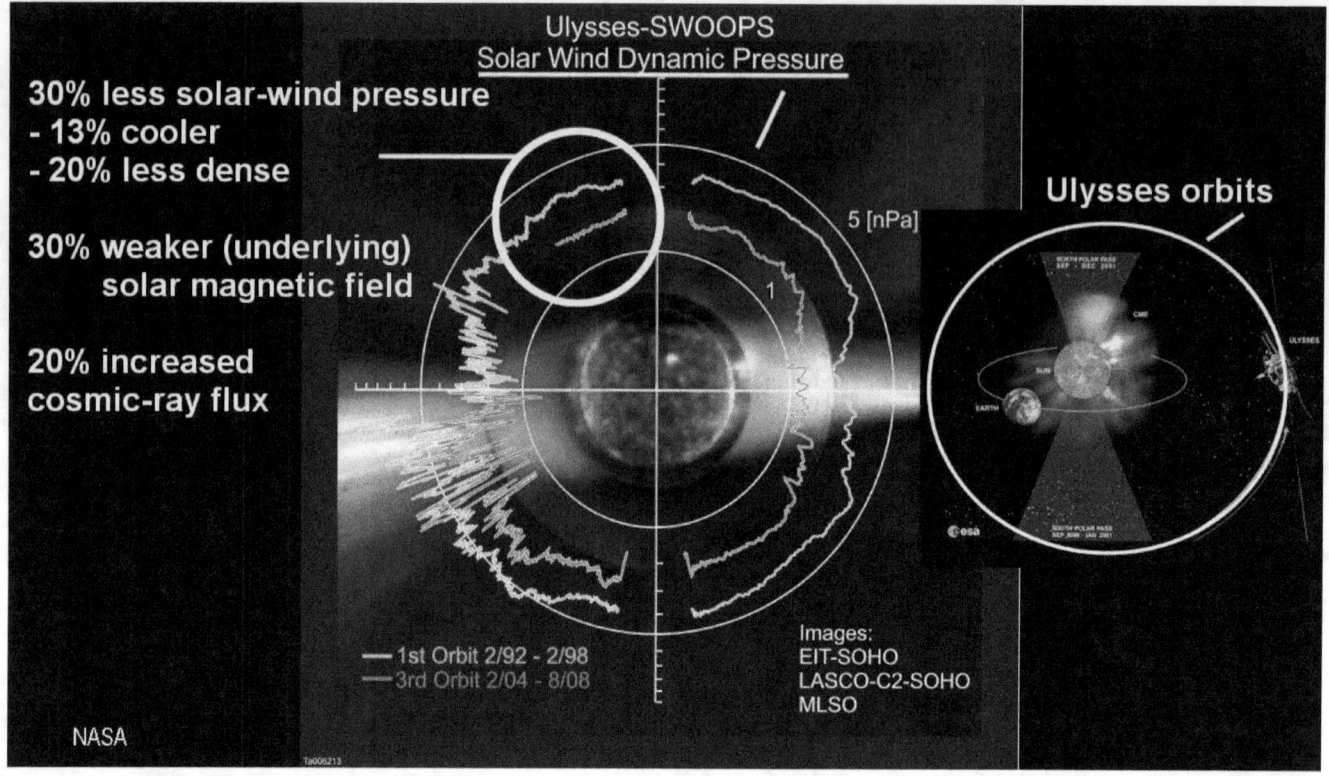

Ulysses saw the same diminishing trend from its very beginning on. Ulysses witnessed the great historic phase shift in solar activity, unfolding.

The solar wind will cease in the 2030s

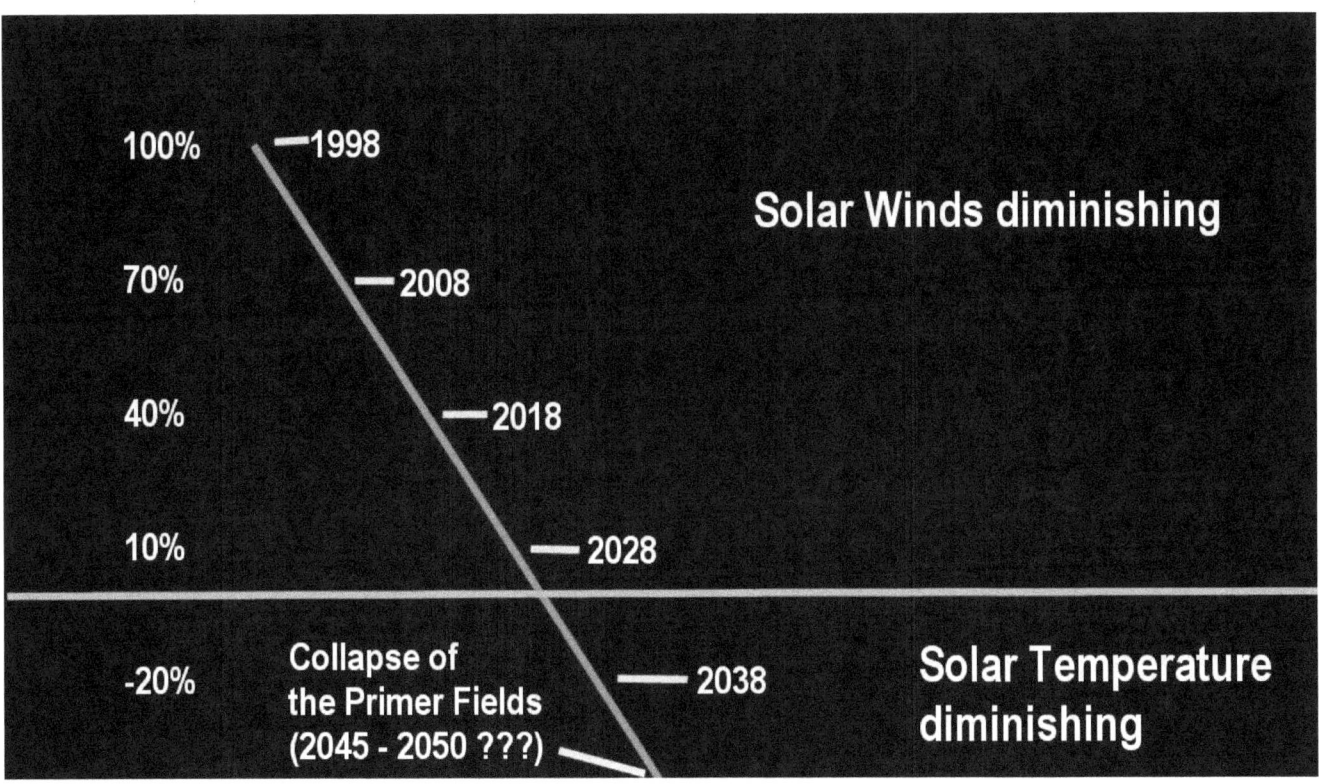

If the collapse of the solar wind continues at the Ulysses-measured rate, the solar wind will cease in the 2030s.

According to the measurements, the solar wind is now at its weakest since the Space Age began.

same rate of weakening, of the Sun

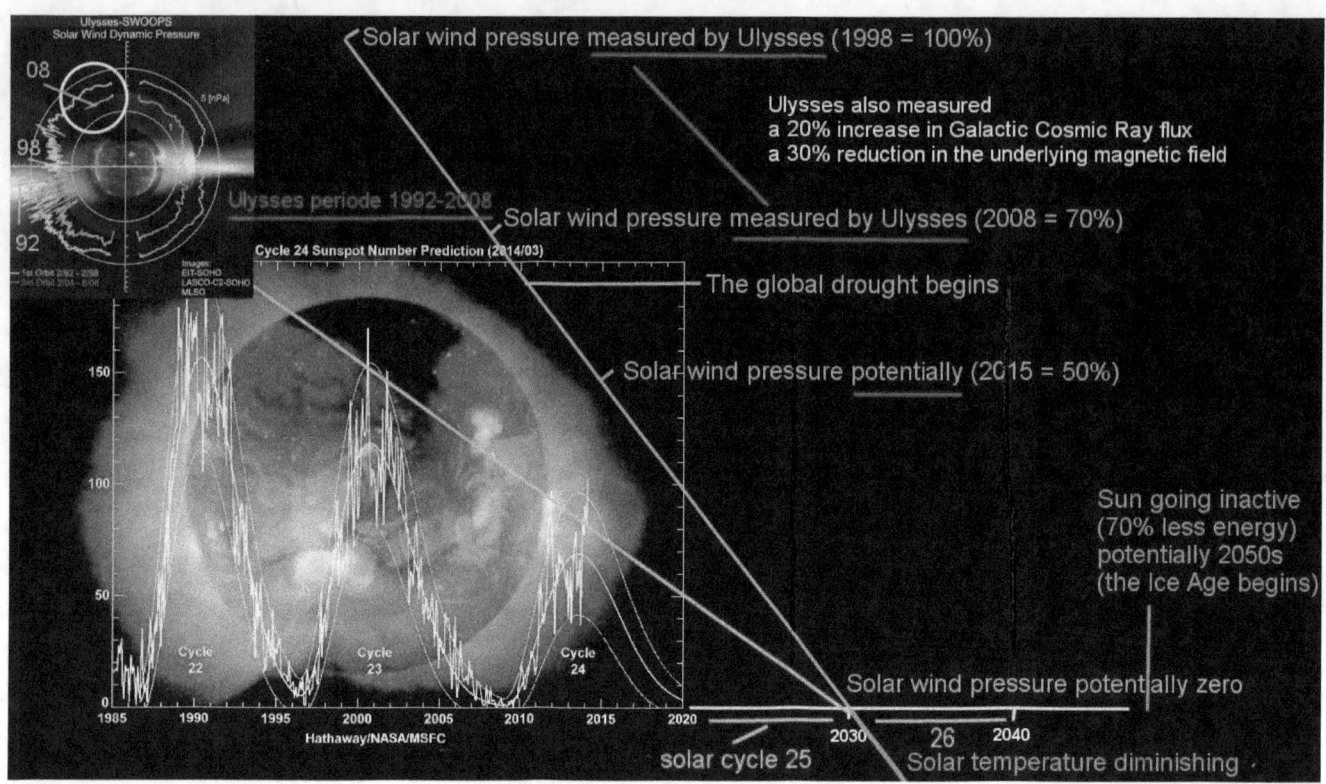

same rate of weakening, of the Sun, that we see evident in the diminishing solar wind, is also evident in the diminishing sunspot cycles. The coincidence with diminishing sunspot cycles confirms that the Sun is indeed weakening.

The end of the solar wind in the 2030s

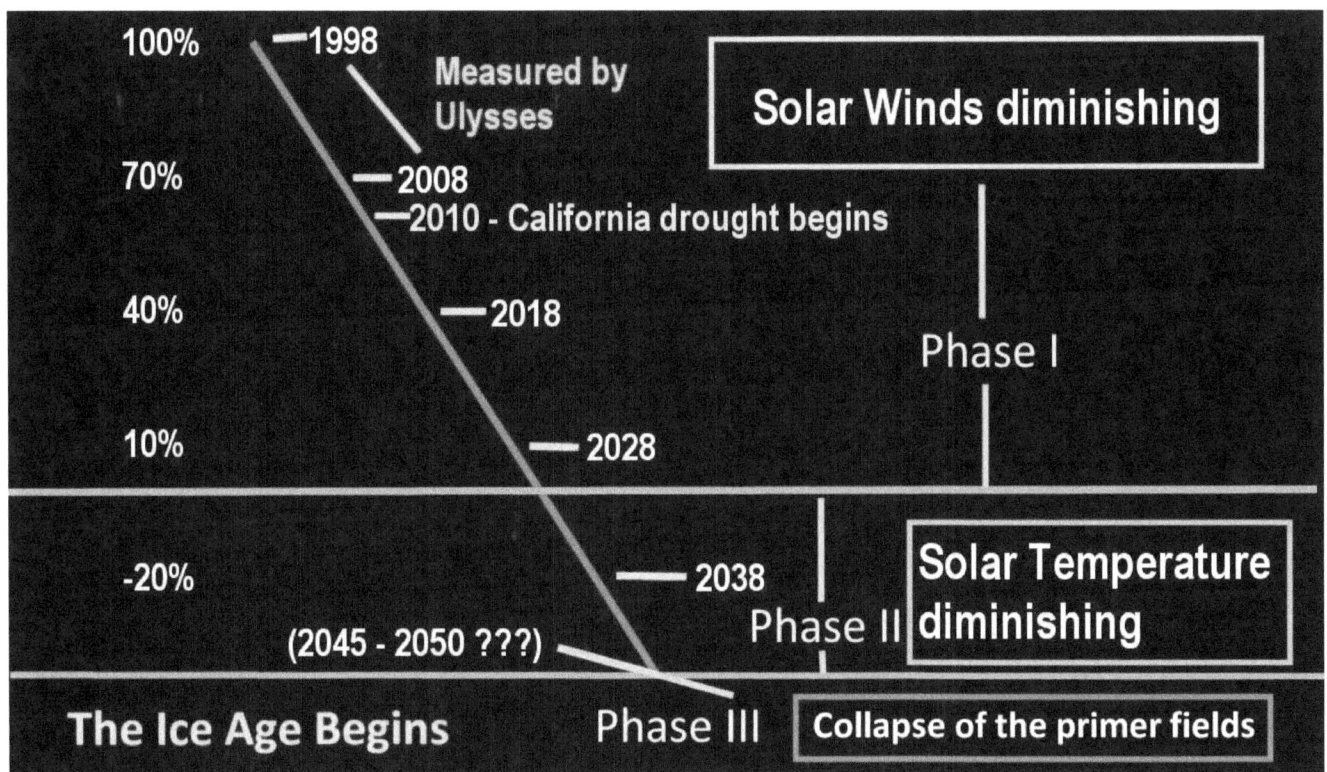

The end of the solar wind in the 2030s will end the first phase of the startup of the next Ice Age, and the start of Phase II in which the Sun's surface temperature begins to diminish. The two phases will most likely overlap to some degree. Climate conditions will arise similar to the little Ice Age. However, there won't be a recovery of the Sun happening this time.

The recovery of the Sun in the 1700s

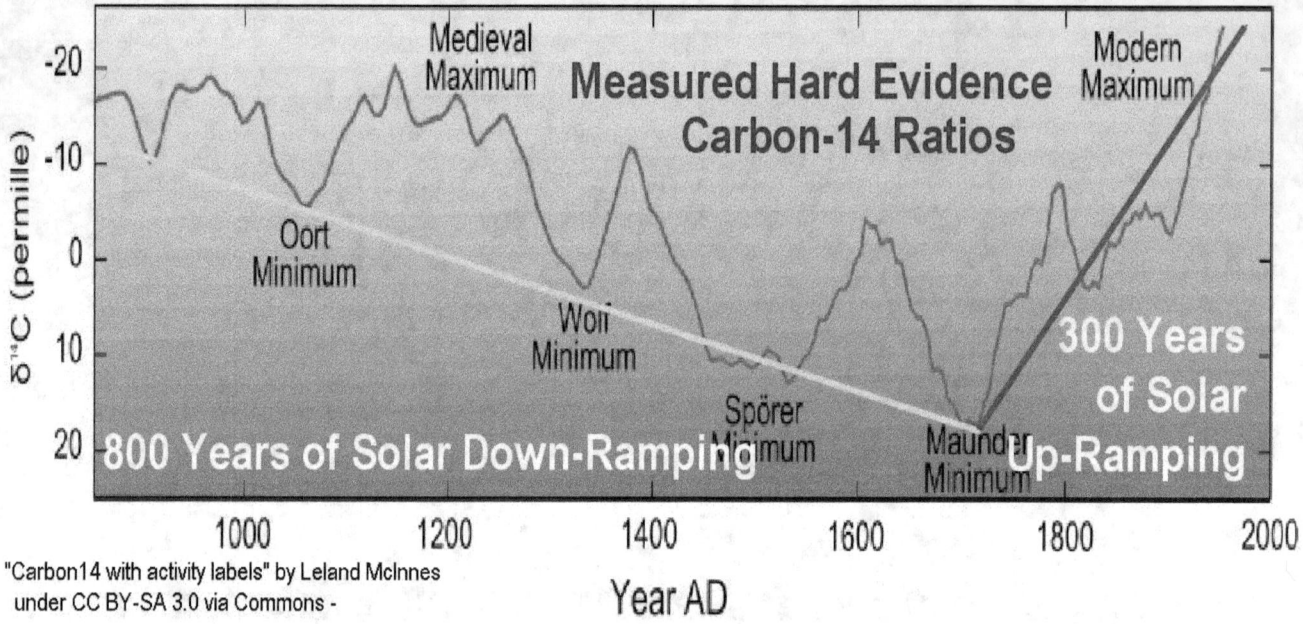

As I said before, the recovery of the Sun in the 1700s falls in line with a series of up-ramping plasma pulses that we see evident in ice core temperature records.

The pulses have been getting weaker for 3,000 years

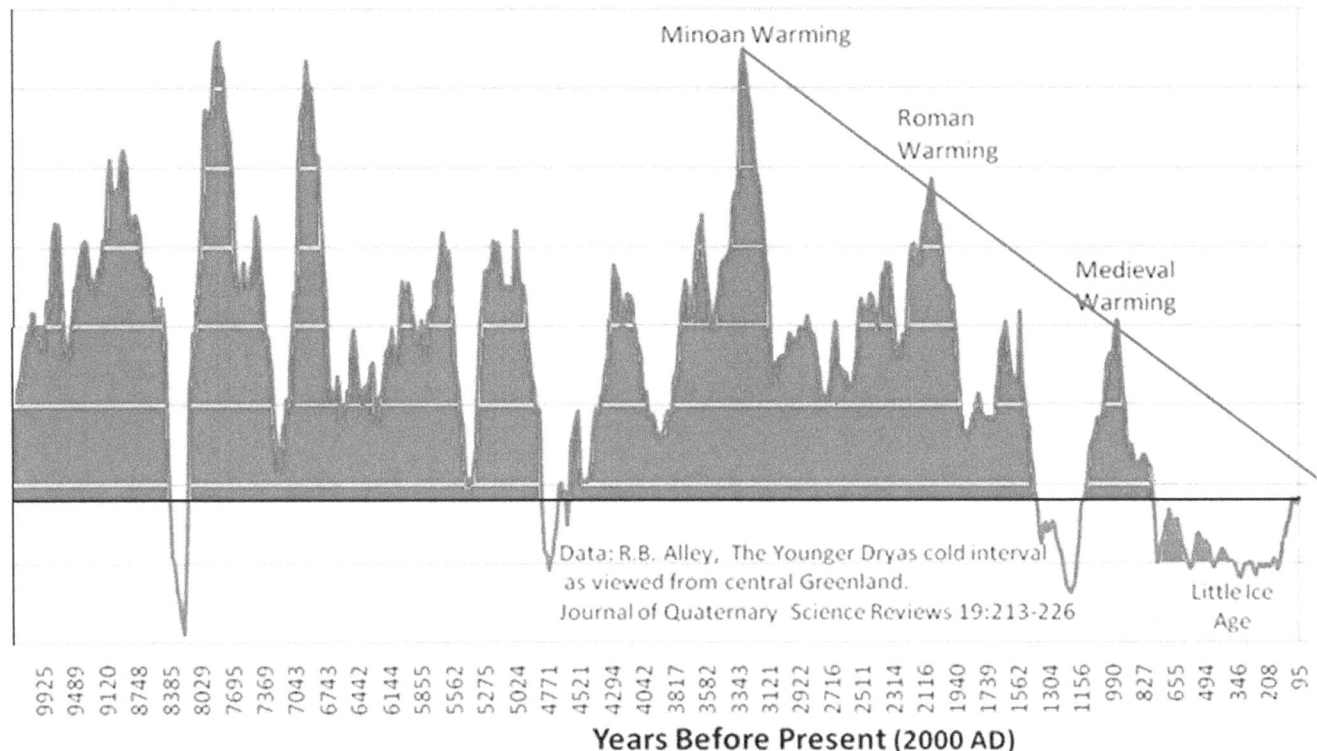

The series of these pulses is such that the next sequential one could have occurred in the 1700s. The pulses have been getting weaker for 3,000 years, and their repetition shorter. If the recovery of the Sun in the 1700s came from this succession, it won't be repeated again for several hundred years. That's far too late to save us from the next Ice Age.

Until a set of the Sun's primer fields collapses

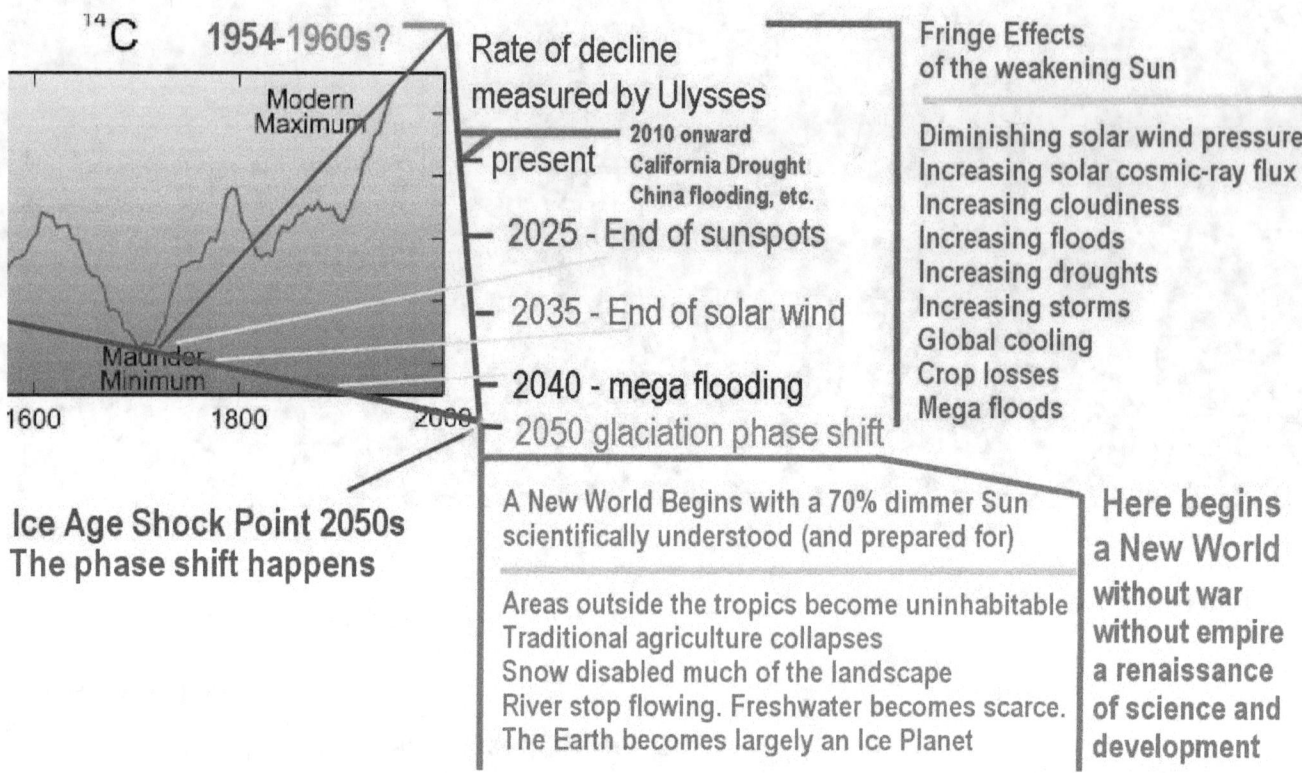

When the plasma stream feeding into the Sun continues to weaken until a set of the Sun's primer fields collapses that focuses plasma onto the Sun, then the Sun will fall back to a low-level default state and will most likely radiate 70% less light and energy. This will potentially happen in the 2050s.

Timing will depend on the resilience of the primer fields

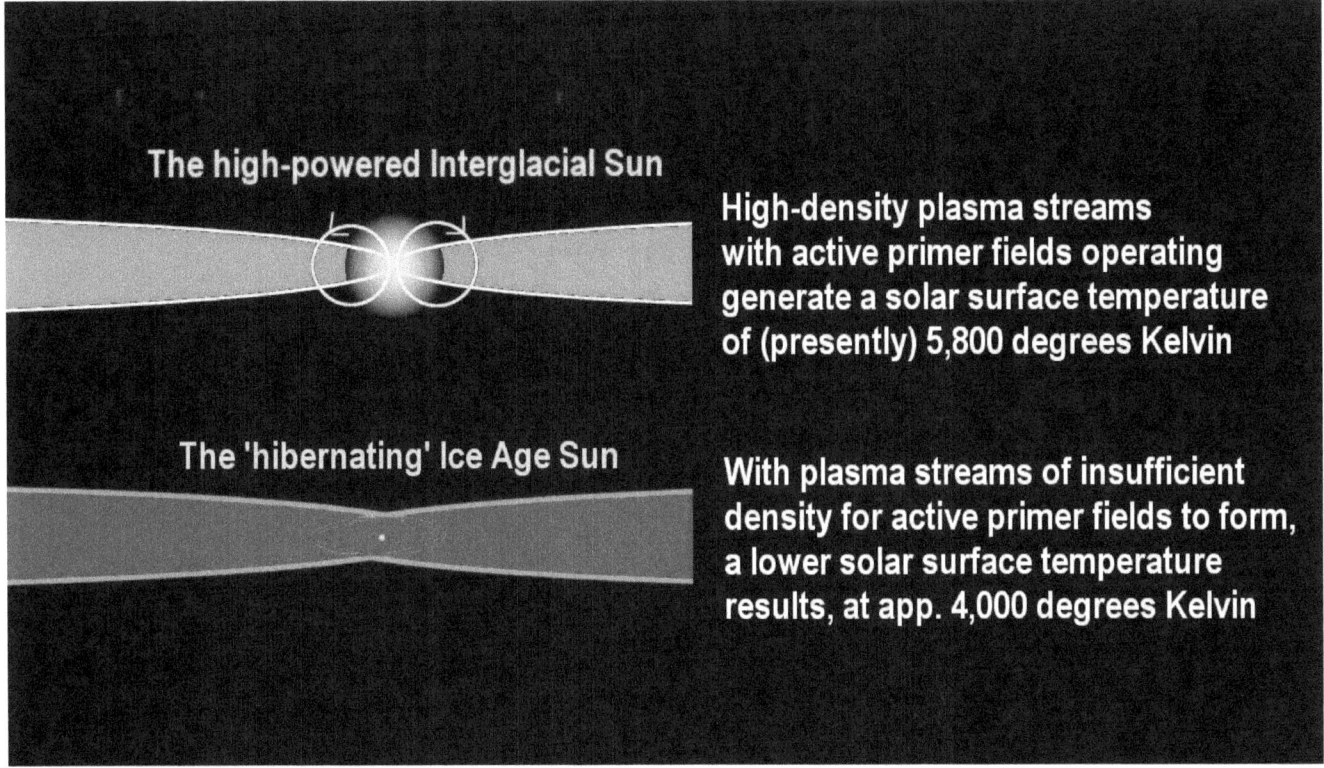

The final timing will depend on the rate of diminishment of the plasma streams feeding the Sun, and the resilience of the primer fields.

Primer fields are electromagnetic structures that are formed by the self-pinching effect of free-flowing electric plasma particles. The particles are magnetically drawn together as they flow, which decreases the cross-section of their path, which in turn increases the rate of its flow. At a point in the progression the magnetic confinement breaks down and allows concentrated plasma to escape, which typically flows onto a Sun that consumes a portion. The weakened plasma stream then flows on in the reverse of the process. But when the plasma stream is too weak, it lacks the density to form a node point. When this happens, it flows thinly around a Sun that consumes a portion and the rest flows on. The transition between the two states may occur in the 2050s or possibly sooner.

As I said before, the active Sun is surrounded by high-density plasma that blocks much of the Sun's cosmic-ray flux. As this Sun gets weaker, its shield gets weaker and its cosmic-ray emissions increase. The emissions will increase until the Ice Age begins.

The inactive Sun, in contrast doesn't have this shield. Its cosmic-ray emissions are therefore whatever the Sun generates at its weak state. The emission diminishes also as the hibernating Sun diminishes.

During the startup of the last Ice Age

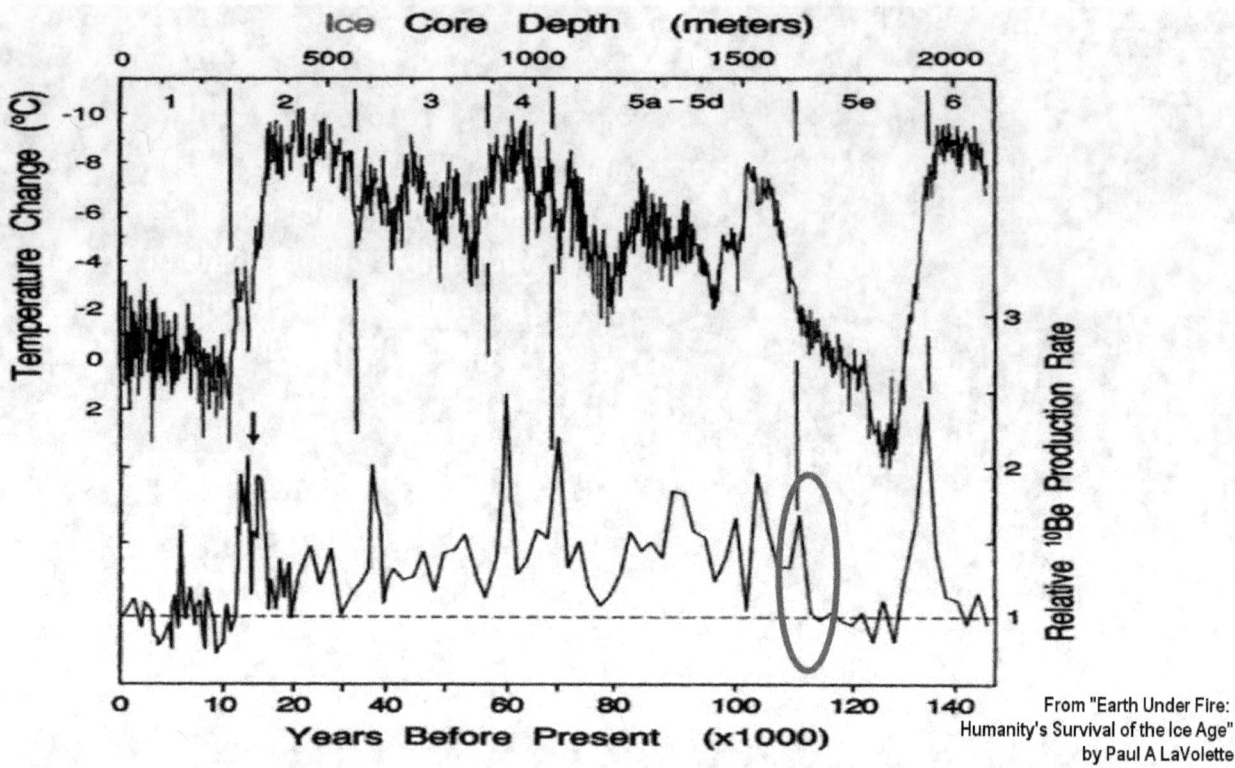

From "Earth Under Fire: Humanity's Survival of the Ice Age" by Paul A LaVolette

During the startup of the last Ice Age, the Beryllium level increased roughly 60% before the glaciation phase shift happened. The Ulysses spacecraft reported a 20% increase of cosmic-ray flux over 10 years. If there exists a linear relationship between changes in cosmic-ray flux and Beryllium levels, then the cosmic-ray increase that Ulysses has reported will bring us to the 60% level in 30 years from the year 2000. This means that the Ice Age phase shift could potentially happen in the 2030s already.

- Are we ready?

Are we ready?

Are we ready?

Are we prepared to live in an uninhabitable world

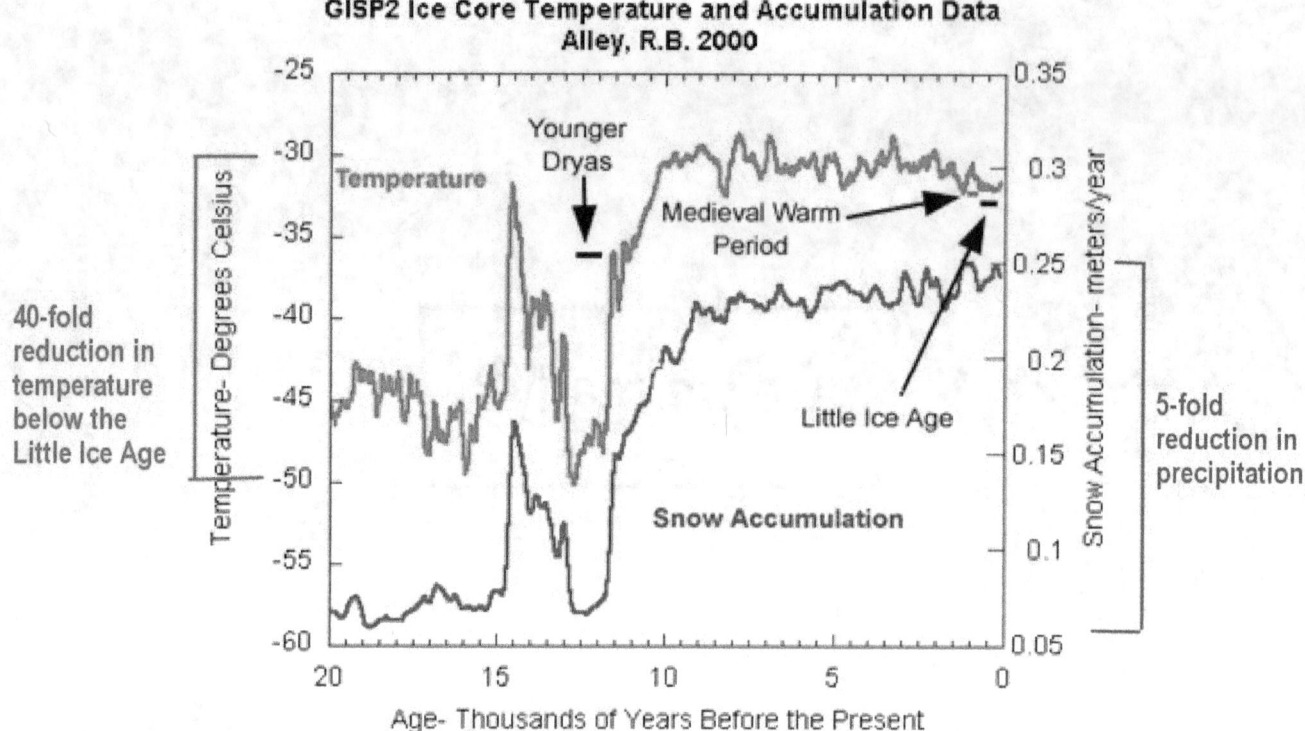

The big question is, are we prepared to live in an uninhabitable world, which the world will largely be when the phase shift happens 30 years from now, with a 40 times deeper cooling than the Little Ice Age had brought, and 75% less precipitation?

The answer to the big question is presently, NO. I see no interest.

Then, do we intend to make the necessary preparations? Again, at the present time the answer is, NO. Society is asleep.

Will we rouse ourselves to meet the Ice Age Challenge, and inspire us with with the determination to win? I would say that the answer is presently a distant "maybe."

Nevertheless, the potential to win on this front does exist. We have the resources to win, and we have the spirit within us to mobilize what it takes to give us a future, if we care to have one. Why then would we fail?

Society remains choked by its limiting perceptions

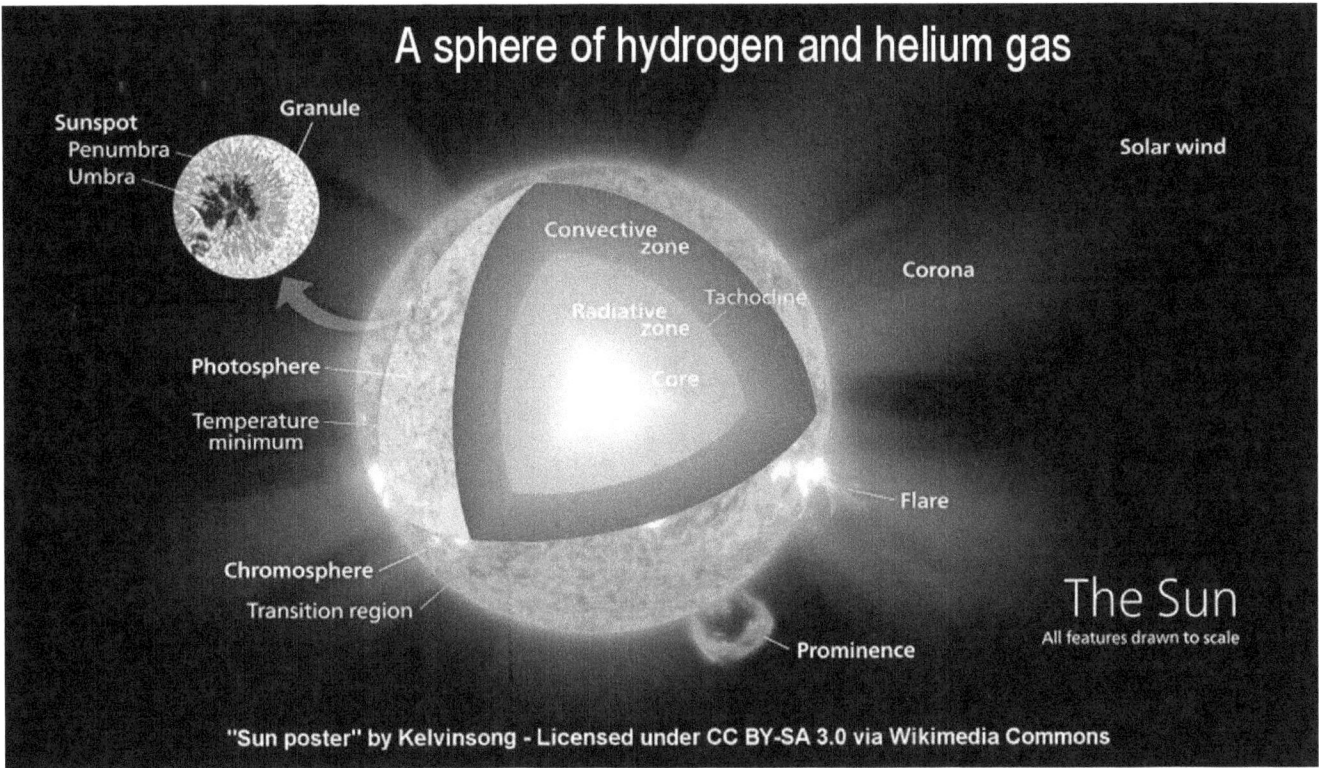

We fail presently, because society remains choked by its limiting perceptions of mechanistic physics, which is a low-level type of physics that is presently the mainstream of physical science

Plasma physics, ABOVE mechanistic physics

Plasma physics, in contrast, operates on a completely different platform than mechanistic physics. It operates on a platform ABOVE mechanistic physics. There, the electromagnetic force and its principles are King. In plasma physics the operating forces are 39 orders of magnitude stronger than gravity and mass, and the related dynamics.

Plasma physics does not operate on the back of mechanistic physics, but operates above it all, on its own platform.

Once the Ice Age Challenge becomes widely recognized

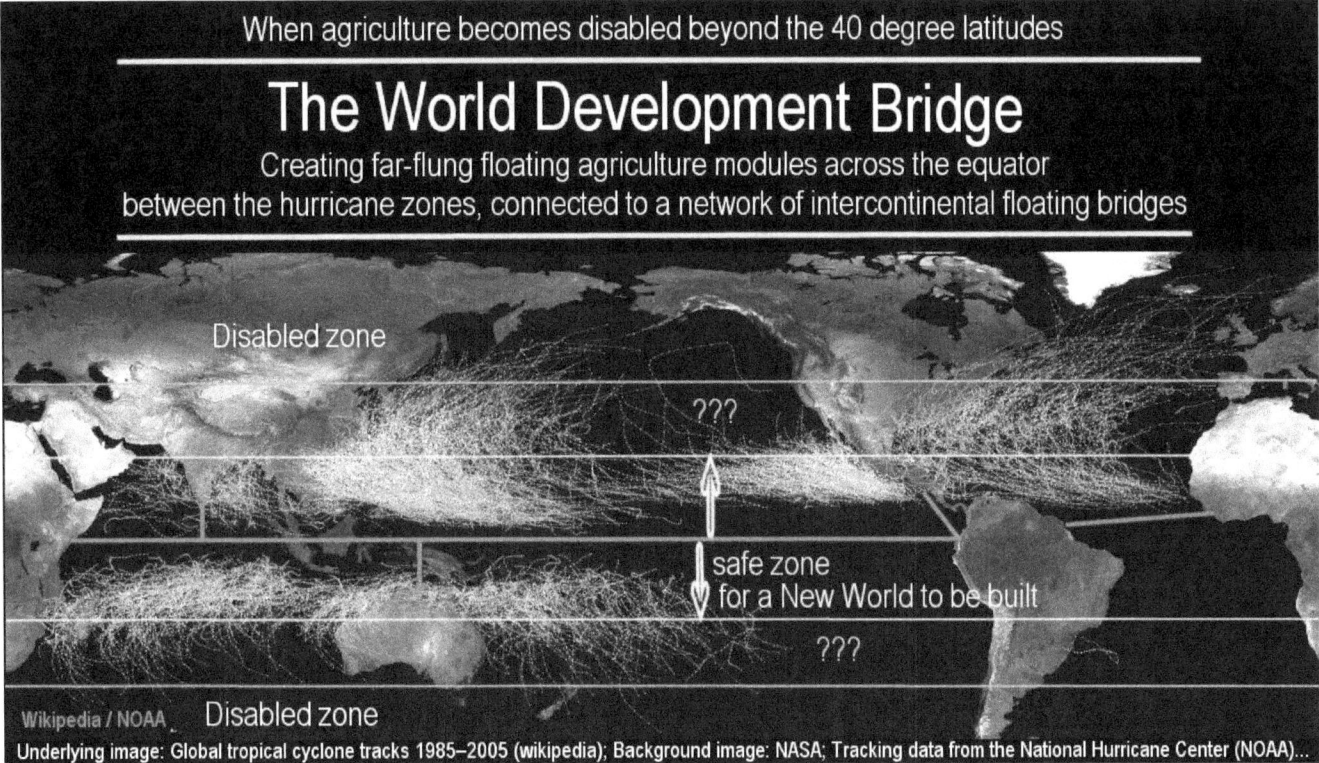

Once the existence of cosmic plasma and its principles is recognized in society, a new world unfolds before it with countless opportunities and with the power in society to prepare itself for the next Ice Age, and protect its existence by meeting its challenges.

The Ice Age Challenge is fundamentally nothing more that a challenge for society to create itself a grand Renaissance. This cannot be done by society remaining asleep in low-level science. Consequently the challenge will cause society to awake.

So I say that we will win, as unlikely as this presently may seem. It is not the nature of the human being to lay itself down to fail at such an existentially critical juncture as the Ice Age startup is, and in the face of the grand opportunities that an awaking to reality offers.

The universe is moving, as it must by the effects of its principles. It revealed its movements, and with this revelation it inspires us to move ahead of it, which we have the capacity to do.

Thus I say, that once the Ice Age Challenge becomes widely recognized and scientifically understood, humanity will mobilize itself to meet that challenge and move ahead of it, and with it mobilize whatever it takes to get this done. And because this challenge is a universal challenge for the whole of humanity, I propose when the challenge is taken up seriously, all the little games of wars, poverty, looting, imperialism, greed, terror, depopulation, and so on, including nationalism, will be laid aside.

On this basis I forecast that we, humanity, have the brightest future ahead of us as the ball towards this future is already rolling, and is rolling amazingly fast, both in the domain of science and in politics, for which the evidence is slowly coming to the surface.

The end

More from the author:

14 Libraries of books and video productions

Novels on Universal Love, the greatest principle in civilization - 14 major novels

Flight Without Limits (science fiction)

Brighter than the Sun (nuclear war avoidance?)

A series of twelve novels: **The Lodging for the Rose**
exploring the Principle of Universal Love

Book 1 - **Discovering Love**

Book 2 - **The Ice Age Challenge**

Book 3 - **Roses at Dawn in an Ice Age World**

Book 4 - **Winning Without Victory**

Book 5 - **Seascapes and Sand**

Book 6 - **The Flat Earth Society**

Book 7 - **Glass Barriers**

Book 8 - **Coffee Sex and Biscuits**

Book 9 - **Endless Horizons**

Book 10 - **Angels of Sex in Queensland**

Book 11 - **Sword of Aquarius**

Book 12 - **Lu Mountain**

The Sex and Sacrament Project - exploration stories from my novels - 11 books

The Son of God

Impotence and Power

      Self-Love and the Golden Hijab

      Erica's Flower Garden

      Helen a Healer

      Brilliance of a Night

      Gem of the Universe

      The Sound of a Bird Woke Me

      Between Ice and Spirit

      Anton of Grace

      Goodness of Living

The Kaleidoscope Project - mixed media of stories from my novels
- videos, PDF, audio

Discovering Infinity - developing history - 13 major research books:
A Research Book Series focused on scientific and spiritual development

    Volume ii (Introduction) **Roots in Universal History** (Focus on Reality)

    Volume 1A **The Disintegration of the World's Financial System** (Focus on Truth)

    Volume 1B **Crimes Against Humanity** (Life Denied)

    Volume 2A **Science and Christian Healing** (History as Truth)

    Volume 2B **The Lord of the Rings' Metaphors**

    Volume 3A **Universal Divine Science: Spiritual Pedagogical** (Structure for Discovery and Scientific Development - The Scientific Process to Know the Truth)

    Volume 3B **Science and Health with Key to the Scriptures in Divine Science**

    Volume 3C **Bible Lessons in Divine Science - 1898**

    Volume 3D **Living in the Sublime**

    Volume 4 **Light Piercing the Heart of Darkness** (The Demands of Truth and Justice)

Volume 5 **Scientific Government and Self-Government** (Platform for Freedom)

Volume 6A **The Infinite Nature of Man** (The Fourth Dimension of Spirit)

Volume 6B **Leadership** (The Spiritual Dimension of Leadership)

Cool Science of Kids - Illustrated Science - **interactive, videos, and 20 books**

War, Economics, and Nuclear War - scientific exploration - **10 videos**

Civilization - series focused on humanity - **10 videos**

Global Warming Doctrine - science videos - **12 videos**

Freshwater and Energy - science videos - **7 videos**

Christian Science explorations - **16 videos**

Books by Mary Baker Eddy - Christian Science - **16 on-line books**

Books by Rolf Witzsche on Christian Science - **9 Books**

The Giant PDF Library all transcripts of videos in PDF form

For links, please see: http://www.ice-age-ahead-iaa.ca

The projects are designed to draw the riches of our humanity into the foreground **towards a New Renaissance**, in order that their light may out-shine the systems of empire that are erroneously accepted, including the follies of war, terror, looting, economic destruction, science-perversion, and policies for depopulation.
Rolf A.F. Witzsche